5/99

The
Science Times
Book of
FOSSILS
AND
EVOLUTION

Other books in the series

The Science Times Book of Birds
The Science Times Book of Fish
The Science Times Book of the Brain

The
Science Times
Book of
FOSSILS
AND
EVOLUTION

EDITED BY

NICHOLAS WADE

THE LYONS PRESS

Printed in the United States of America

Designed by Joel Friedlander, Marin Bookworks

10 9 8 7 6 5 4 3 2 1

Library of Congress Cataloging-in-Publication Data

The Science times book of fossils and evolution / edited by Nicholas Wade.
      p.   cm.
    ISBN 1-55821-652-9
    1. Fossils. 2. Evolution (Biology) I. Wade, Nicholas.
  II. Science times.
  QE723.S48   1998
  560—dc21   97-46955
                                                      CIP

Contents

Introduction

NOTHING IS AS DEAD as a fossil, yet the study of fossils is as lively as any branch of science. As the raw material for understanding the history of evolution, fossils have never ceased to amaze and provoke.

To hold a fossilized ammonite or shark's tooth in your hands is to witness the death of a creature that died tens or hundreds of millions of years ago, an event that transcends historical time. You may wonder at the beauty of this structure built of stone, or at the accident by which this one individual among the vanished millions of its kind left a permanent trace of its existence.

Fossilization, in fact, requires a series of accidents. The animal must die and somehow become hastily entombed in mud or sand or silt before scavengers can dismember it. This natural shroud must then lie undisturbed for years while the slow processes of mineralization take place, whether replacing the tissues or making a cast that later sediments will fill.

Then this body-turned-to-stone must wait for eons until the rock around it erodes away, exposing it to a fossil hunter's gaze.

For much of human history, fossils were objects of complete bafflement. In the Middle Ages they were dismissed as a *delusus naturae*—"joke of nature." The idea that they were exactly what they seemed to be, the relics of long-dead creatures, proved extraordinarily hard to accept.

Only after geologists like Charles Lyell and James Hutton had laid the basis for supposing the Earth was far more ancient than the few thousand years implied in the biblical account, and after Darwin had proposed his theory of evolution, was the intellectual framework created in which fossils could be understood for what they were.

Then as now, they were usually at the center of some fierce controversy. This was in part because the theory of evolution challenged theologians' long-accepted answers to questions about the origin of life and the universe. Also, fossils provide a very partial sample of the plant or animal communities from which they are drawn and often raise more questions than they answer.

In the last century the furious debates about evolution between bishops and biologists were an argument over how to interpret the fossil evidence. The embers of that clash still smolder among creationists, who prefer the account of origins given in Genesis to that of the theory of evolution.

Today's arguments are more about the details of evolution than its overall framework. Yet these details include some fundamental questions. Three issues have been of particular prominence in the last few years.

One is the origin of life. Though scientists have sometimes despaired of ever penetrating the mists that surround this distant event, a new approach to the problem has generated hope. The new idea is that life started not in the familiar surroundings of a lake or tidal pool but in conditions of extreme heat such as are seen in hot springs or the volcanic vents that gash the deep ocean floor.

Both these environments are now known to teem with life, including microbes whose origins are as ancient as any on the planet. This and other evidence hint that the natural chemicals from which the first living systems emerged may have been ingredients of a steaming-hot brew, some hellish cauldron near a volcano or ocean vent, and that life only later adapted to the far cooler temperatures at which it is now prominent.

The idea has given a new focus to origin-of-life studies, which had reached something of a dead end. Fossils give a minimum age for the beginning of life on Earth, but the earliest moments of life occur in one of the fossil record's large blank spaces. Biologists, however, can get some help in exploring this desert from a method that is increasingly used to supplement the fossil record, that of studying the messages embodied in the genetic code.

These messages are written in the language of DNA, the molecule whose string of chemical letters spells out the instructions for creating and operating every living organism. When the DNA sequences of equivalent genes from different organisms are compared, it is often possible to draw up a family tree showing how the genes are related to each other.

The DNA letters suffer accidental changes or mutations over the course of time, a process that is a driving force of evolution. It is also the basis for estimating the age of the DNA, since some categories of mutation occur at a fairly regular rate and so serve as what are known as molecular clocks.

Molecular clocks have proved useful in a second issue of perennial interest, the origin of humans. Fossils of human remains have been all too scarce. Only recently, after molecular clocks indicated the age of the best rocks to explore, have early human fossils started to appear in reasonable numbers. Two new species of hominid have recently been found that lived right at the branching point between the ape and the human lines. They may in time shed light on the causes of this decisive event.

A third major issue concerns the causes of the mass extinctions that have occurred several times in the Earth's history. Paleontologists have long believed that these puzzling extinctions should be explained in terms of natural processes such as sea level changes and volcanic eruptions. But at least one of these extinctions, that which rang down the curtain on the Cretaceous period, has now been shown beyond doubt to have been caused by an extraterrestrial agent, a large asteroid or comet that slammed into the Yucatán Peninsula 65 million years ago.

Paleontologists are working hard to find out how many other extinctions may have similar causes. When their work is done, the history of this planet may assume a quite different texture. Far from being a peaceful blue sphere immune from the violent physical forces that shape the universe, the Earth may turn out to have been periodically devastated by catastrophic impacts that wiped out life on land and sea.

The articles selected for this book first appeared in the Science section of the New York Times. The illustrations were conceived by the section's art directors, Nancy Sterngold and Michael Valenti.

Journalism is like the May fly, that labors long hours as an unseen grub, then appears for a day and dies. My colleagues on the Science section and I are grateful to Lilly Golden and to Nick Lyons of the Lyons Press, who have given our ephemeral words the chance to live a little longer than usual, an extension of life that the ever-vital subject of fossils surely merits.

—NICHOLAS WADE, Spring 1998

1

HOW LIFE BEGAN

The problem of how life began is among the outstanding mysteries of science.

No traces of the first life have yet been found in Earth's rocks. There is no information from other planets to tell how often, if ever, life has originated elsewhere in the universe.

Despite the lack of clues, scientists studying the origin of life have made considerable progress in the last few years. They still cannot reconstruct the first faltering chemical steps that spanned the watershed between non-living and living. But they have gathered much new information about the context in which this event may have occurred. A burst of new ideas has encouraged many biologists to renew the assault on a long-intractable problem.

One new finding is that the window of time in which life must have emerged is much smaller than had been assumed. It used to be thought that there were a billion years or more for life to evolve from the chemicals present on the primitive Earth. But scientists now think the planet would have been inhospitable to life for hundreds of millions of years after its formation some 4.6 billion years ago. Chunks of leftover space debris surrounding the young Sun would have continued to bombard the Earth, and the larger of these objects generated enough punch to vaporize the primitive oceans.

These titanic impacts are sometimes referred to as "sterilizing events" because they would probably have wiped the Earth clean of any incipient life forms. The bombardment did not cease until 3.8 billion years ago.

Meanwhile, the fossil history of life on Earth has been pushed back to far earlier starting dates. Cell-like objects as old as 3.5 billion years have been discovered, and there are also reports of chemical traces indicative of life in rocks that are 3.85 billion years old.

If these dates are accurate, life on Earth had to evolve with remarkable rapidity. And if so, then life, far from being an improbable event requiring a billion years of chemical dice-throwing, may be a rather likely event, given favorable conditions.

Another new insight relates to the nature of those conditions. In the past, biologists studying the origin of life have tended to assume that it got started in environments similar to those today, perhaps after a few lightning strikes created suitable chemicals in a warm pool. But recently bacteria and other organisms have been found in hot springs and deep-sea vents, living at temperatures above the boiling point of water.

These microbes have so ancient an ancestry that they point to the possibility that life originated at high and hellish temperatures, adapting later to cooler conditions. If so, the origin of life should be sought in a place quite different from the biologists' favorite candidate spots: not in tidal pools or warm ponds but deep below the surface of the Earth, whether in deep, hot rocks or at the undersea vents where underwater volcanoes erupt through the ocean floor.

The two new concepts—that life gets started quickly, and its origin may be deep and hot—put a quite different focus on the search for life beyond Earth. The other planets and moons of the solar system, which once seemed definitely barren, now look as if they may hold many promising habitats. There is suddenly a more urgent case for scouring the solar system for signs of life.

————————————

Fossils Prove Life Came About Faster

FOSSIL MICROORGANISMS discovered in Australia by an American scientist prove that life was already thriving and diversified 3.485 billion years ago, leaving a much narrower window than previously thought for life to develop. The microorganisms, representing 11 distinct types, are more than 1.3 billion years older than any comparable fossil group ever found. This dramatically curtails the time available for life to have evolved naturally on Earth, and could focus more attention on the disputed hypothesis that life originated elsewhere in the universe and somehow reached Earth from afar.

At the least, the new findings are likely to prompt a reexamination of rates at which species are presumed to have evolved on the early Earth.

In a report in the journal *Science,* Dr. J. William Schopf of the University of California at Los Angeles describes 11 distinct types of microorganisms found in an unusual rock formation in northwestern Australia. The microscopic fossils were embedded in tiny mineral grains, which were encased in another type of rock that scientists were able to determine formed 3.485 billion years ago.

The grains in which the fossils were found are almost certainly older than the rock encasing them, but Dr. Schopf has been unable to determine just how old that might be.

All the Australian fossils are of single-celled organisms in the form of microscopic filaments less than a hundredth of an inch long. Their worm-like shapes, Dr. Schopf said, resemble those of certain modern bacteria and cyanobacteria—the most primitive organisms capable of photosynthesis.

In photosynthesis, the process that fuels most plants, the energy of sunlight is used in synthesizing carbohydrates and other complex biological building materials from the carbon dioxide in the planet's air and water.

These pioneers of life on Earth may have founded the lineage that many eons later led to both plants and animals, including humans. Because of their similarity to modern cyanobacteria, Dr. Schopf believes, the organisms may already have developed photosynthesis as a source of fuel—a surprising achievement for organisms that had so little time to evolve.

Estimates of the ages of the Earth vary, but many scientists believe the planet solidified in its present form about 4.6 billion years ago.

Scientists know that the moon, which formed at the same time, was peppered by devastating meteor impacts during the infancy of the solar system, and the Earth must have been subjected to the same bombardment. Even if life began during this early period, it would have been quickly extinguished by the fiery impact of meteors.

Many geologists believe that life could not have achieved a permanent toehold on the Earth until the meteor bombardment subsided about 3.9 billion years ago. Hitherto, the dating of the earliest known life form allowed theorists more than one and a half billion years in which to account for the presumable development of living cells from the soup of chemicals covering the primordial Earth. The new finding shrinks that window to a mere 500-million-year time span.

"It may seem astonishing that life could have reached such a degree of complexity and diversity so long ago," Dr. Schopf said in an interview, "but I don't think biologists will be surprised for long."

Most scientists believe that terrestrial life originated on the Earth itself, arising from a primordial soup of amino acids and other carbon-based chemicals by some yet-to-be-explained mechanism. But a few scientists have suggested that life may have drifted to the Earth in the form of spores or single-celled organisms from some other place in the galaxy or universe.

If life began elsewhere, it would not have required a long period for microorganisms to evolve like those found in Australian rocks, and fairly advanced organisms might have appeared soon after the Earth cooled down.

Among those who have proposed the "panspermia" theory, that life arrived on Earth in the form of interstellar microorganisms, are Sir Fred Hoyle, a British theorist, Dr. Francis Crick of the Salk Institute in San Diego, co-discoverer of the molecular shape of DNA, and Dr. Leslie Orgel, a British chemist also of the Salk Institute.

But Dr. Orgel said in an interview that it was not necessary to invoke the idea of panspermia to account for the very early appearance of microorganisms on Earth suggested by the recent discoveries in Australia.

"I'm not really surprised by Dr. Schopf's discovery, nor am I convinced that it necessarily takes a long time for such organisms to develop after life begins," Dr. Orgel said. "Also, I've never insisted that life began elsewhere. It was only a suggestion that might be considered."

Dr. Crick has suggested that intelligent extraterrestrial beings, knowing that life has great difficulty getting started from simple chemicals, might have deliberately seeded the galaxy with such durable organisms as spores or bacteria capable of surviving eons of space travel in suspended animation.

Very old rock is difficult to find, Dr. Schopf said, because most of it is uplifted into mountain ranges and then eroded away over the ages. Any fossils it might have contained are destroyed by this process.

Rocks dating from the Archean eon, or early Precambrian period, are extremely rare, and authenticated biological fossils were almost unknown until the recent discovery. But Dr. Schopf believes that the near-total absence of life in the fossil record during the Earth's early period does not mean that there was no life, but merely that most fossils were destroyed by geological processes.

—MALCOLM W. BROWNE, April 1993

Clues to Fiery Origin of Life Sought in Hothouse Microbes

DARWIN PICTURED LIFE arising in a warm little pond. But scientists are now concluding that the beginning, four billion years ago, was in conditions more like hell and that the progenitor of all terrestrial life was right at home in the searing heat.

Most surprising, experts are finding that today's earthly hot spots teem with organisms that are evolutionary throwbacks to that era. Invisible without a microscope, these tiny specks are now being seen as living relics of the first life, making them as close as science is likely to get to the ultimate ancestor of humans and all creatures.

Once thought of as bizarre oddities, these heat-loving microbes at the outer limits of life have now vaulted to a central scientific role, with the Federal Government devoting more than $10 million to mapping their genes to better understand their workings and evolutionary history. Meanwhile, scientists around the globe are racing to find new ones in terrestrial hot springs and geysers and volcanic vents beneath the sea.

The tiny creatures are known as thermophiles, or heat lovers. In recent years, biologists have discovered them thriving at extraordinarily high temperatures—up to 235 degrees Fahrenheit, hotter than the usual temperature of boiling water.

Some are bacteria. Some are something stranger. In general, they cannot thrive in the absence of high temperatures, quickly going into a dormant state if things cool down too much.

Remarkably, scientists are finding their genes to be ancient, putting them among the deepest roots of the evolutionary tree.

"At this stage of the game, it's a given that the first life was some kind of thermophile," Dr. Norman R. Pace, a microbiologist then at Indiana University who is a pioneer in the study of the unusual microbes, said in an interview.

The origin of life is one of science's most daunting mysteries. Until recently, the conditions under which it arose were a matter of almost pure speculation. Two quite separate lines of inquiry have now edged scientists toward the idea that the Earth's first organisms emerged and lived at very high temperatures.

One is the estimate that the surface of the early Earth probably remained very hot for many eons, into the period when life must have got its first foothold on the planet. The other is the gathering realization that the descendants of this steamy epoch have not only survived but still inhabit special niches reminiscent of their fiery home all over the globe.

Indeed, the heat-loving organisms turn out to be surprisingly common among the many branches of the microbe family, bolstering the idea that they are very old in terms of evolutionary history.

"You find thermophiles in lots of different groups and environments, which suggests that the property of thermophily is a primitive one," said Dr. Thomas D. Brock, a thermopile pioneer at the University of Wisconsin in Madison. "If it were a late adaptation, there'd be no way for it to get around so widely."

The first days of the Earth were hellish, with vast chunks of rubble left over from the creation of the solar system slamming into the planet and helping to keep it hot. The main fusillade hit from 4.6 billion to 3.8 billion years ago, the planet's earliest days.

As the molten Earth cooled a bit, widespread volcanism began to disperse water vapor and other gases, producing a thick atmosphere and vast oceans. Radioactivity was high. A steady fall of cometary ice, some scientists say, also helped to inundate the world and perhaps sowed organic molecules associated with life. A few scientists speculate that perhaps tiny spores of life from elsewhere in the universe may have fallen to Earth to seed the process of biological evolution.

By one alchemy or another, life began, with scientists now putting its start as far back as 4.2 billion years ago, quite soon after the Earth's formation. Intolerable as were the conditions in this infernal hothouse, it must nonetheless have served as the womb of life, rather than the room-temperature conditions often assumed by early theorists.

Whether indigenous or alien, the first life faced the threat of quick extinction, new research suggests. It appears that the rocky bombardment around four billion years ago was still so violent that the entire planetary ocean was probably vaporized repeatedly.

The thickened atmosphere from such gargantuan jolts would have trapped so much sunlight that the Earth's surface temperature became an inferno of more than 3,000 degrees Fahrenheit for thousands of years, sterilizing the Earth's surface.

"It seems likely that the result would have been to 'reset the clock' for life's origins," Christopher F. Chyba, Tobias C. Owen and Wing H. Ip, who are planetary scientists, wrote in *Hazards Due to Comets and Asteroids,* published by the University of Arizona.

As the bombardment slowly let up, temperatures are thought to have settled down, perhaps hovering in the neighborhood of 212 degrees, the temperature of boiling water at sea level. The Earth failed to cool off quickly because the high levels of carbon dioxide in the atmosphere kept trapping lots of solar radiation.

By about 3.8 billion years ago, researchers say, the temperature might have dropped to a balmy 120 to 190 degrees.

The earliest probable fossils, dating to about 3.5 billion years ago, have been found by J. William Schopf of the University of California at Los Angeles and other researchers. These appear to be copies of today's cyanobacteria, thermophiles that live comfortably at 160 degrees Fahrenheit or even higher temperatures.

Pasteur announced in 1864 that heat could kill microbes and sterilize a solid or liquid. Pasteurization, widely adopted over the decades, was a boon to public health and food preparation. But exceptions to its usual workings took a century to discover.

In 1964, Dr. Brock, then at Indiana University, visited Yellowstone National Park, home of the celebrated geyser known as Old Faithful. He was looking for uncommon microbes. While sampling hot waters in a woodsy, out-of-the-way area, alert for grizzly bears, he was surprised to find bacteria thriving at temperatures that were thought to be anathema to life.

From a pinkish mass in 1966, he isolated one that flourished at 158 degrees Fahrenheit—at the time, an unheard-of temperature for life. In

1967, he found microbes living in boiling water. "We kept pushing to higher and higher temperatures," Dr. Brock recalled.

As scientists rushed to global hot spots and uncovered dozens of new kinds of heat-loving microbes, Dr. Carl R. Woese of the University of Illinois in 1977 stunned the scientific world by announcing that some of them constituted a third form of life.

Up to then, scientists had divided the living world into two kingdoms according to genetic packaging. The blueprint of life, DNA, floated free in the cells of bacteria, also known as prokaryotes. In the cells of eukaryotes, like fungi, plants and animals, it was gathered into a nucleus.

The DNA of Dr. Woese's new kingdom floated free but was genetically distinct from the bacteria, extravagantly so. "It was a total surprise," he recalled. "The other surprising thing was the resemblance to the eukaryotes. It seemed like a new group in its own right."

He named the new kingdom archaea (pronounced ar-KEY-a), or ancient ones, since their evolutionary history appeared to link them to the Earth's earliest life.

Knowledge about the global distribution of such strange microbes jumped after the 1977 discovery in the Pacific's icy depths of hot springs, which were later found to be widely distributed in the deep and to swarm with thermophiles.

It jumped further in the 1980's as Dr. Pace took a clever step to speed their collection from global hot spots. Thermophiles are notoriously hard to culture and grow, with only one species in a thousand or so surviving a laboratory's artificial environment. So Dr. Pace skipped that step. He pioneered the direct analysis of DNA from field samples.

The power of the method was recently driven home by Dr. Susan M. Barns, a colleague of Dr. Pace at Indiana, who pulled genetic evidence of more than 40 different types of hot microbes from Yellowstone's Obsidian Pool, which is deep black in color and boiling hot. Preliminary analysis shows that two of its archaea may be among the most primitive organisms on Earth.

"It's in the back country," Dr. Barns said of the pool. "There's thousands of hot springs back there."

Of late, scientists have also found thermophiles living deep under the Earth itself, where temperatures are uniformly high. Some have been isolated from hot oil reservoirs beneath the North Sea and Alaska's North Slope. Others have been retrieved from depths nearly two miles beneath the continental United States, living at temperatures as high as 167 degrees.

Most important, genetic analyses that highlight the evolutionary relationships among creatures have revealed these heat-loving organisms to be the most primitive on Earth, literally living fossils from the earliest days.

The main way of drawing such evolutionary trees is to look at RNA, which contains genetic information copied from the genes. A particularly good kind of RNA to examine is found in the ribosomes that make proteins inside cells. By comparing the ribosomal RNA of different organisms, researchers can discover degrees of evolutionary relatedness.

Based on such analyses involving more than 2,000 organisms, scientists believe that the archaea and bacteria diverged from a common ancestor four billion years ago, soon after life arose. Only later did the forerunners of today's eukaryotes split off from the archaeal branch of the evolutionary tree.

In addition to shedding light on evolution, thermophiles are turning out to be a treasure trove for the biotechnology industry. So great is the overall interest that the Federal Government is supporting four projects to map their DNA. Blueprints for RNA make up less than 1 percent of the genetic material in the typical archaea, so the expansion of the analysis to include the DNA is expected to produce a rush of insights.

No scientist has yet announced the mapping of an organism's complete DNA sequence. But archaea are good candidates because they have relatively small amounts of genetic material, simplifying the analysis. A human cell has about three billion base pairs of DNA. An archaea has about two million.

DNA mapping is now under way on three different kinds of archaea, financed by the Energy and Commerce Departments. Some results are already coming in and the work is to be finished in 1998.

"This initiative will expand our knowledge of how life has evolved and how it sustains itself," Dr. Martha Krebs, a senior official of the Energy Department, said when its Microbial Genome Initiative was announced last year.

Groups now mapping DNA are the University of Utah in Salt Lake City, the Institute for Genomic Research in Gaithersburg, Maryland, the University of Illinois, the Genome Therapeutics Corporation of Waltham, Massachusetts, and Ohio State University.

Knowledge of the complete DNA sequence will let companies clone bits of it, including the genetic blueprints for enzymes, which are key proteins for performing cellular rearrangements.

Companies that have mapped some thermophile DNA on their own are racing to isolate, clone and sell the extremely heat-stable enzymes for use in genetic engineering. These enzymes can speed the polymerase chain reaction, or PCR, which has spawned a branch of biotechnology that allows scientists to rapidly make trillions of copies of vanishingly small amounts of DNA, including the DNA in fingerprints at crime scenes.

But many scientists say the greatest payoff from the mapping work will be enhanced understanding of the origins of life. Dr. Woese says the federally financed DNA work is already buttressing and fine-tuning the phylogenetic maps based on ribosomal RNA analysis.

Ultimately, he added, such work is likely to bolster the idea that life began in an inferno, not a warm pond. "The evidence suggests that both the archaea and the bacteria are of thermophylic ancestry," he said. "And that makes a strong case that their common ancestor, which is the ancestor of all living things, was a thermophile."

—WILLIAM J. BROAD, May 1995

Evidence Puts Date for Life's Origin Back Millions of Years

A TEAM OF AMERICAN, British and Australian scientists has uncovered evidence that life may have originated on Earth at least 350 million years earlier than previously believed, at a time immediately after a devastating bombardment by meteors.

The discovery complicates the question of when and how terrestrial life arose. Recent studies based partly on the distribution of craters on the moon have suggested that a rain of meteors smashing into the early Earth was intense enough until 3.8 billion years ago to destroy all life on the planet, if any existed at that time. The oldest objects believed to be the fossils of living organisms—tiny mineral filaments—are estimated to be 3.5 billion years old.

But findings reported in the journal *Nature* imply that the window of time between the end of the lethal meteor bombardment of the young Earth and the beginning of life may have been shockingly short. The work suggests that life had already established a toehold on the infant planet 3.85 billion years ago.

In a comment also published by *Nature*, Dr. John M. Hayes of the Woods Hole Oceanographic Institution in Massachusetts said the investigation seemed to imply that biochemical processes "developed with breathtaking rapidity after the last large impact."

Many scientists believe that the chemical processes that gave rise to life must have taken hundreds of millions of years to develop the essential enzymes, proteins and genetic codes. Evidence of a very quick beginning might mean that there was too little time to complete the job on Earth, and that life therefore must have originated elsewhere, drifting through space to the Earth in the form of spores or by some other means. This is known as the panspermia theory.

The discovery was by a group headed by Dr. Gustaf Arrhenius of the Scripps Institution of Oceanography at San Diego. (Dr. Arrhenius is the grandson of Svante Arrhenius, the renowned 19th-century Swedish scientist whose work laid much of the basis of modern chemistry and who believed that life originated somewhere other than the Earth.) The team also included Dr. Arrhenius's graduate student Stephen J. Mojzsis and colleagues at the University of California at Los Angeles, the Australian National University in Canberra and Oxford Brooks University in England.

The focus of their investigation was a formation known as the Isua supracrustal belt of Akilia Island in southern West Greenland. The formation had been dated by radioactive decay techniques at 3.85 billion years old, the oldest known rocks in the world.

The discovery did not include any actual fossils. It is believed that no fossils could survive from such ancient times. Fossils can be preserved in sediments for hundreds of millions of years, but eventually all sediments and the fossils embedded in them are drawn under the sliding tectonic plates that make up the Earth's crust, or "subducted." When this happens, the sedimentary fossil record is crushed, melted and destroyed.

But the scientists in Dr. Arrhenius's group hit upon another way to search for ancient life.

The carbon atoms in molecules distributed around the world consist mostly of an isotope, or form, in which the carbon nucleus contains six protons and six neutrons. This is called carbon-12. But among these atoms are small numbers of another isotope, carbon-13, whose nucleus contains six protons and seven neutrons.

When environmental carbon from carbon dioxide and other substances is incorporated into living organisms, a process of "fractionation" occurs in which the heavier isotope, carbon-13, is somewhat reduced. A reduced ratio of carbon-13 to carbon-12 is therefore a characteristic of biological activity, and Dr. Arrhenius and his group said they believed this changed ratio had been preserved even in minerals that had undergone "pressure cooking" by geological processes.

By analyzing tiny particles of carbon embedded in some of the Greenland formation's apatite rock, a phosphate mineral chemically similar to the apatite in bone, the group found a ratio of carbon-13 to carbon-12 that the scientists believed revealed the signature of life.

To study the specks of carbon, the group used a high-resolution ion microprobe, which fires a beam of atomic nuclei at the sample. The beam blasts atoms loose from the sample's surface, which are then analyzed using an instrument called a mass spectrometer. The scientists found that their samples fell into two groups: those of clearly non-biological origin, and those displaying an isotope ratio characteristic of life.

Dr. Arrhenius said in an interview that "no terrestrial process other than biological organisms has been found that fractionates carbon-12 and carbon-13 in this way."

Whether or not the premises on which this work is based are true remains controversial, however. "There are some potential processes other than life that could account for the Arrhenius isotopic results," Dr. Hayes of the Woods Hole Institution said, "and even though they are long shots, we cannot entirely discount them. Still, if I had to rate his paper as a whole, I'd say it boils down to good science."

Dr. Stanley L. Miller of the University of California at San Diego, whose experiments beginning in the 1950's suggested that some essential components of life could have been created by the interaction of simple chemicals in "the primordial soup," believes that the narrowing of the window in which life arose presents no problem.

"There's nothing to rule out a quick beginning of life," he said in an interview. "Life might have arisen in, say, 10 million years, or even less. Or, it may be that the first wave of life was wiped out by the late meteor bombardment, except for organisms that thrive on heat, like those we find today living around thermal vents on the ocean floor. Something like them may have been our ancestors. There are lots of possibilities."

—MALCOLM W. BROWNE, November 1996

The Comet's Gift: Hints of How Earth Came to Life

A TRILLION or so comets are thought to lurk beyond the planets on the dark fringes of the solar system. Lost to a realm where the Sun's faint rays do nothing to lessen the interstellar chill, most of these whirling chunks of dirty ice orbit endlessly.

But over recorded time, a few comets have left this deep freeze and sped into the inner solar system, at times lighting Earth's skies. Past civilizations often saw them as harbingers of death and doom, and irrational reactions to comets still occur, as evidenced by the grim suicides in Rancho Santa Fe, California, where 39 people died in the apparent belief that Hale-Bopp was their ticket to aliens and extraterrestrial bliss.

Modern astronomers are fascinated by the comet Hale-Bopp for other reasons. Never before have they witnessed anything quite so spectacular. Its icy core is estimated at 25 miles wide, more than 10 times the size of the average comet and big enough to swallow many Manhattans. Its great size makes it unusually bright and easy to study. Since it was first discovered beyond the orbit of Jupiter, astronomers have scrutinized it unceasingly as the Sun warmed its outer layers, causing the gargantuan ice ball to shed many tons of clues every second about the nature of its chemical makeup.

Now the first comprehensive findings are in and give support to a remarkable theory. It suggests that cometary ices bear the chemical precursors of life and that comets fell on the aboriginal Earth in vast numbers and sowed these precursors for what eventually became the planet's riot of biological diversity. The same mechanism is thought to be at work throughout the cosmos, sowing the seeds of life on untold worlds.

This view is now getting major support as telescopes around the globe find Hale-Bopp spewing not just tons of water but methanol, formaldehyde,

carbon monoxide, hydrogen cyanide, hydrogen sulfide and many compounds rich in carbon—in other words, the basic ingredients thought to be
necessary for the origin of life.

"This is the ironclad link to the new paradigm," Dr. Dale P. Cruikshank,
an astronomer at the National Aeronautics and Space Administration's Ames
Research Center in California, said in an interview. "We've never had such
a panorama of important molecules."

Dr. Cruikshank is the first author in a series of eight articles on the
Hale-Bopp findings in the journal *Science*. The new discoveries are seen as
a milestone in the developing field of bioastronomy, which looks to the heav-
ens for the chemical forerunners of life. And the tidings of cosmic fertility
are strikingly at odds with the old view of comets as portents of doom.

Hurtling through space at more than 27 miles a second, the comet is
visible for months throughout its orbit. Such a comet appears "once every
200 years or so," said Dr. Brian G. Marsden of the Harvard-Smithsonian
Center for Astrophysics in Cambridge, Massachusetts, who helps track
comets and other heavenly bodies for the International Astronomical Union.
"What makes it great is its persistence."

Dr. Harold A. Weaver, an astrophysicist at Johns Hopkins University
and lead author on one of the *Science* papers, said Hale-Bopp might actually
outperform itself while heading back toward its icy abode.

"The nucleus sometimes stores heat and might remain at elevated temperatures for longer outbound than inbound," he said in an interview. "We
already have this beautiful set of data and we want to watch it go back out
again. There's clearly a lot of excitement."

Some telescopes must stop tracking Hale-Bopp as it approaches the
Sun or be damaged by bright sunlight. This is particularly true for the Hub-
ble Space Telescope orbiting Earth. Later, as the comet races toward the dim
hinterlands, astronomers will redouble their efforts to gather clues to Hale-
Bopp's chemical and physical makeup in hope of strengthening the uncanny
link between comets and the first stirrings of life on Earth.

Such thinking is a radical departure from the traditional view developed over the decades by scientists, mainly geologists and geophysicists.

In the old picture, Earth was thought to have coalesced out of primordial dust as a bare sphere with no atmosphere, basically a rocky desert. The
gases and water vapors and carbonaceous brews that formed the atmosphere

and filled the seas were seen as having come from within Earth in an early period of intense volcanism. Lightning storms then stirred the primordial soup and created carbon-rich molecules that organized themselves into self-replicating units, or crude forms of life.

Not so, says a newer theory, which astronomers, astrophysicists and planetary scientists tend to advance.

More than four billion years ago in its early days, Earth was hot enough to expel into space most of the water and lighter materials and chemicals, this theory holds. So the planet remained barren rock.

For the makings of life, the new theory looks to the wastelands out among the stars, especially to the dark clouds that loom against starry backdrops. Such dim zones are common among the countless stars that make up the Milky Way, Earth's galaxy. On a clear dark night, the naked eye can easily see the voids amid the milky brightness.

These dark interstellar clouds turn out to be peppered with grains of matter about the size of talcum particles that are virtual factories for the production of complex chemicals. To date, scientists have identified nearly 100 molecular species.

Interstellar water was found in 1968, formaldehyde in 1969, methyl alcohol in 1970, hydrogen sulfide in 1972, methylamine in 1974, ketene in 1976, methane in 1978, ethylene in 1980, methyl diacetylene in 1984, methyl isocyanide in 1987 and butatrienylidene in 1990. Such molecules were discovered to make up about half of the interstellar matter, the rest being mainly atomic hydrogen.

These chemical finds were made as astronomers turned radio, infrared, ultraviolet and visible-light telescopes on the icy wastelands and studied signals that showed molecules giving off their own precise signatures amid a chaos of electromagnetic waves.

One pioneer of such analysis is Dr. Yvonne J. Pendleton of the Ames Research Center. She works there with Dr. Cruikshank, her husband and occasional co-author. In the March 1994 issue of *Sky and Telescope* magazine, the couple wrote about finding evidence of chemical precursors to life in space.

No signs of amino acids—the building blocks of proteins—have yet been found amid the interstellar wastes. But lots of forerunners have been found, including ammonia and hydrogen cyanide. The leap to amino acids is said to take only the addition of liquid water.

Today, scientists see the interstellar factory as working in several steps. First, dying stars lace the void with an array of relatively simple molecules such as methane (CH_4), water (H_2O) and carbon monoxide (CO), as well as tiny grains of silicon, the element found in rocks, sand and semiconductors. Over eons, the molecules coat the silicon grains with icy mantles.

Dense clouds of such materials condense in places to form new stars that irradiate nearby grains with bursts of ultraviolet light, transforming the simple molecules into more complex ones like formaldehyde (H_2CO) and methyl alcohol (CH_3OH).

In theory, such complex interstellar dusts then become the raw material for new generations of stars, perhaps accompanied by planets as well as trillions of icy comets.

Astronomers on Earth observe comets rushing through the inner solar system only fleetingly. In contrast, the interstellar wilds are always visible. Despite the difficulty, astronomers, aided by recent advances in telescopes and instrumentation, have begun to learn some of the chemical secrets of the icy visitors.

The first glimmers came in 1985 and 1986 with Halley's comet, which yielded signs of water, carbon dioxide and formaldehyde (used on Earth as a disinfectant and preservative). In 1989, another comet showed evidence of methyl alcohol (used as antifreeze), while one in 1990 gave signs of hydrogen sulfide (the poison that smells like rotten eggs).

The big advance came last year with comet Hyakutake, which was extremely bright and easy to analyze. Its signals gave evidence of many different molecules akin to those of the interstellar wastes.

Now astronomers studying Hale-Bopp have topped even that. The findings are significant both for the sheer number of molecules and, as important, for a detailed description of when the chemicals evaporated from the giant ice ball as it raced inward from the deep freeze toward the Sun.

In *Science*, a 12-member team headed by Dr. Nicolas Biver at the Observatoire de Paris-Meudon, in Meudon, France, describes nine chemicals sighted by radiotelescopes: carbon monoxide (CO) in September 1995, methyl alcohol in March 1996, hydrogen cyanide (HCN) in April 1996, hydroxyl radical (OH) in April 1996, hydrogen sulfide (H_2S) in May 1996, formaldehyde in June 1996, carbon monosulfide (CS) in June 1996, methyl cyanide (CH_3CN) in August 1996 and hydrogen isocyanide (HNC) in November 1996.

"It's a landmark," Dr. Cruikshank said of the list. "It shows when and how these molecules turn on, and shows the evaporation sequences. They're also found in space. This is the link that really clinches the connection to the interstellar medium and proves that this material is essentially unaltered."

Put differently, it shows that comets are heavenly vans carrying tons of carbon-rich materials from interstellar space to addresses throughout the solar system.

Dr. Cruikshank said bombardments of cometary ice undoubtedly hit not only the early Earth (heavily for a period of perhaps a billion years or so) but also Mercury and the moon. Both those bodies have recently yielded signs of hidden ice.

Even today, he added, microscopic dust particles from comets and rocky asteroids rain down on Earth at the rate of about 300 tons a year.

The prevailing wisdom, bolstered by Hale-Bopp, is now that comets played a pivotal role in begetting life on Earth.

"Thirty years ago we thought you had to have lightning storms," Dr. Michael J. Mumma, a longtime comet expert at NASA's Goddard Space Flight Center in Maryland, said in an interview. "Now it's recognized that the materials are delivered automatically."

"That doesn't mean you get life out of that," Dr. Mumma added. "But it means you deliver large amounts of simple chemicals, and maybe some complex ones, that can lead to life directly."

—WILLIAM J. BROAD, April 1997

Evidence Backs Theory Linking Origins of Life to Volcanoes

THE THEORY that life on Earth began around a volcano, perhaps at the deep-sea vents where molten lava boils through the ocean floor, has been bolstered by the chemical reconstruction of an essential step in the metabolism of living cells.

If the new finding is correct, it means that the recipe for creating life on a newborn planet consists of mostly lethal ingredients and would read something like this: Drop a handful of fool's gold (the mineral iron pyrites) and a sprinkle of nickel into water, stir in a strong whiff of rotten eggs (caused by the gas hydrogen sulfide) and carbon monoxide, heat mixture near the crackle and hiss of a volcano and let simmer for an eon.

Some kind of natural chemical reactions presumably preceded the emergence of the first living cells sometime before 3.5 billion years ago.

But the nature of these reactions, a subject known as pre-biotic chemistry, has become a matter of dispute as the textbook theory, developed by Dr. Stanley L. Miller of the University of California at San Diego, has come under challenge.

The latest effort to reconstruct pre-biotic chemistry has been made by Dr. Claudia Huber of the Technical University of Munich and Dr. Gunter Wachtershauser, also of Munich. Dr. Wachtershauser is a Ph.D. in chemistry who works as a patent lawyer but is respected among chemists who study the origin of life. Their article appears in the journal *Science*.

Dr. Christian de Duve, a biochemist and Nobel Prize winner who has written on the origin of life, said Dr. Wachtershauser's new work was "an extremely interesting finding which fits with the idea that life may have originated in a volcanic setting."

"It stresses the importance of sulfur and iron, which again fits with what we know from biochemistry," he said.

Dr. Robert H. Crabtree, a Yale University expert on metals and chemical change, said it was tough to imagine that anyone would discover exactly how life started "but nevertheless this is an important contribution" and one that "could come to be seen as comparable" with Dr. Miller's if it should prove correct.

There is a growing belief among some microbiologists that the locale for the origin of life was not tidal pools and lightning strikes, as implied in Dr. Miller's view, but in deep vents or other geothermal sources. "We are seeing a mini paradigm shift," said Dr. Norman R. Pace, a microbiologist at the University of California at Berkeley. "It's a field-wide split and Wachtershauser has certainly been a driver."

The classic experiment performed by Dr. Miller when he was a graduate student in 1953 seemed to solve the scientific problem of how life began, at least in broad outline. He took some water to represent the ocean, the gases methane, ammonia and hydrogen to represent the early Earth's atmosphere, and sent electric sparks through the mixture to simulate lightning strikes. After several days he found that many organic chemicals typical of living cells, including amino acids, the building blocks of proteins, had formed within the concoction. It seemed only a matter of time for chemists to figure out how the building blocks might have combined naturally into the complex molecules of life.

But taking the next steps beyond Dr. Miller's brilliant beginning proved much harder than expected. One reason is that geologists changed their minds about the composition of the Earth's early atmosphere. They now think it consisted largely of carbon dioxide and nitrogen, a far less reactive mixture than Dr. Miller used.

"I would call the conventional origin of life chemistry as being at a dead end," said Dr. Carl R. Woese, a microbiologist at the University of Illinois.

It was Dr. Woese who discovered that all living creatures belong to three very ancient lineages: bacteria, eukarya and archaea. The archaea, single-celled organisms, are of interest to biologists studying the origin of life because they have certain primitive features and a liking for extreme environments, like the boiling springs of Yellowstone National Park and the superheated waters that swirl from deep-sea volcanoes.

Deep-sea vents, first directly observed only 20 years ago, have become of increasing interest as a possible origin of the planet's life. These underwater volcanoes, or black smokers, are home to a strange array of creatures found nowhere else, as well as to many species of archaea. The vents spew out superheated water laden with a black cloud of chemicals, including carbon monoxide, hydrogen sulfide and various metal sulfides.

These are the chemicals that have been studied by Dr. Wachtershauser, whose theory is that many reactions important to the origin of life took place on the surface of ferrous sulfide, fool's gold. In his latest experiment he has shown that with a slurry of ferrous and nickel sulfides to help the reaction, the gases known to be present in black smokers will form acetic acid and an activated form of acetic acid known as a thio ester.

The synthesis of these chemicals is striking to biochemists because it demonstrates a natural way of chaining carbon atoms together, the basis of the living cell's chemical repertoire. The thio esters are also akin to a vital component, known as acetyl co-enzyme A, of the energy metabolism pathway known as the citric acid cycle.

Dr. Woese is enthusiastic about Dr. Wachtershauser's work, saying, "His theory in my opinion is the only game in town because it is predictive and therefore testable."

He emphasized the theory's implication that the first forms of life created their own vital chemicals instead of sucking them in from the outside, as implied in the Miller version.

Dr. de Duve, from his knowledge of biochemistry, has long believed that a sulfur-based chemistry involving thio esters was the probable basis from which the RNA world arose. The RNA world is the concept that ribonucleic acid, or RNA, which still plays many essential roles in living cells, was the predecessor of DNA as the repository of hereditary information.

Iron-sulfur clusters exist at the center of many important enzymes, Dr. Pace noted.

"They are fundamental biology and were invented back in the very early days," he said. "Wachtershauser has dreamed up a number of wonderful iron-sulfur chemistries that could be involved."

Dr. Pace believes that the primary biochemistry of life was acquired in a geothermal sludge before the invention of the cell membrane. "It wasn't cells crawling out of that sludge, it was something deeper," he said.

Dr. de Duve, who has joint appointments at Rockefeller University in New York and the University of Louvain in Belgium, said from his home in Belgium that Dr. Wachtershauser "disagrees with Stanley Miller and vice versa" but that there were likely to be important elements of truth in both theories.

Dr. Wachtershauser said that as an amateur he had held off publishing his theory for fear of being ridiculed but was encouraged by the late philosopher Karl Popper and others to go ahead. The "soup theory," he said in reference to Dr. Miller's version of the origin of life, had produced only inactive chemicals whereas his own theory, based on the idea of chemicals being held on a two-dimensional surface, showed how a basic metabolic cycle could start and be maintained.

"Once I had published," Dr. Wachtershauser said, "I was received with open arms."

Dr. Miller did not respond to a telephone message seeking comment.

While not accepting the whole system, Dr. de Duve said Dr. Wachtershauser's latest results agreed with the general hypothesis "that life may have originated in a hot, deep-sea environment rich in sulfur."

If so, such a dark, hellish cradle would be a considerable elaboration on the "warm little pond" in which Darwin, in an often-quoted letter, suggested that life began.

—NICHOLAS WADE, April 1997

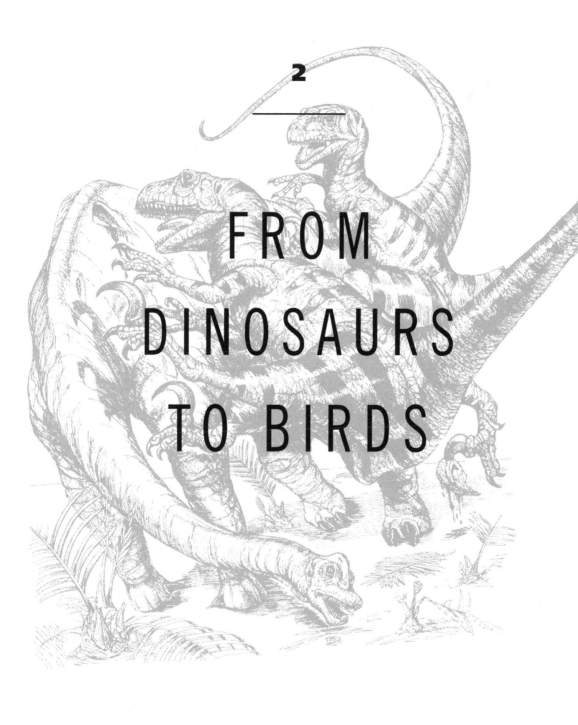

2
_____

FROM
DINOSAURS
TO BIRDS

Though life seems to have arisen on Earth at almost the earliest possible opportunity, it was in the form of single cells. The next step in evolution, the emergence of multicellular forms of life, was apparently much harder, since it took about three billion years. Once nature had mastered this difficult trick, an enormous proliferation of life forms occurred in what is called the Cambrian Explosion of 540 million years ago.

The long gestation period before the Cambrian is known to paleontologists as the Precambrian era. Though it seems a distant, shadowy era, it accounts for three-quarters of the planet's history.

The Cambrian Explosion opened a 300-million-year experiment known as the Paleozoic era. Marine invertebrates like the trilobites spread throughout the oceans. The first fishes evolved. Both plants and animals emerged from the sea and learned to live in that strange, specialized environment known as dry land. First amphibians flourished, then the reptiles. But just as the experiment was going full blast, a catastrophe brought it to a sudden end. Numerous species on land and sea fell victim to the great Permian extinction.

The Paleozoic era was over, but the horrendous extinction cleared the stage for the exuberant drama of the Mesozoic era. During the three scenes of its 185-million-year duration, the Triassic, the Jurassic and the Cretaceous, the dinosaurs rose, flourished and fell.

These extraordinary creatures continue to arouse fear and fascination. Fresh discoveries are made every year as paleontologists find new fossils and reinterpret old ones. The field is full of lively debates as biologists explore new aspects of the creatures' lives, from the noises they made to the diseases that afflicted them.

Still unresolved is how and where among the dinosaur family the birds developed, the single form in which these once ubiquitous creatures escaped extinction.

————————————————

Michael Rothman

Two utahraptors, or "superslashers," attacking a brachiosaurid. Superslashers weighed about a ton and could reach a length of about 20 feet. Each hind foot had a 15-inch upright slashing claw.

Fossil of a Terror
That Dwarfed *T. Rex* Is Found

EXPLORING THE SAHARA in southeastern Morocco, paleontologists have found the fossil remains of two huge predatory dinosaurs. The discovery is seen as a major step in uncovering Africa's early fossil past and understanding the evolutionary changes in dinosaurs after the breakup of the continents.

The more spectacular of the finds is the gigantic skull and sharp teeth of a meat-eating creature that lived 90 million years ago and measured 45 feet from its snout to the tip of its tail. The skull, with a length of five feet four inches, may be larger than the largest skull of a *Tyrannosaurus rex,* which lived 70 million years ago in North America and had long been considered the largest known terrestrial carnivore.

But this African dinosaur apparently was not as bright as tyrannosaurs; its brain cavity appeared to have half the volume of the later and more familiar predator. Scientists identified the fossil skull as belonging to a species first recognized in 1927 but poorly known until now: *Carcharodontosaurus saharicus,* which means "shark-toothed reptile from the Sahara."

The other discovery is the partial skeleton of a predatory dinosaur completely new to science. As reconstructed in the laboratory, the dinosaur is at least 25 feet long and, for a creature so large, has unusually long, slender limbs, suggesting that it was a swift and agile hunter. The fossil animal is named *Deltadromeus agilis,* or "agile delta runner."

"Nothing like *Deltadromeus* has been found on any other continent so far," Dr. Paul C. Sereno, a paleontologist at the University of Chicago who led the expedition that made the discoveries last year, said.

By the time these two predators lived in Africa some 90 million years ago, Earth's land had become a patchwork of isolated continents. The single supercontinent in existence when dinosaurs first appeared 230 million

33

A Creature to Make *T. Rex* Tremble

Scientists have recently discovered the bones of several specimens of a species of dinosaur billed as the deadliest land creature the Earth has seen, which they have named utahraptor. The animal, nicknamed superslasher, weighed about one ton, reached a length of about 20 feet, was a swift runner and was armed with a 15-inch slashing claw that stood upright and apart from the other claws on each hind foot.

The animal's forelegs had large paws tipped with powerful claws suitable for grasping prey while the dinosaur kicked its victim with its sickle-clawed hind feet. The utahraptor was described by its finders as a "Ginsu-knife-pawed kick-boxer" that could disembowel a much larger dinosaur with a single kick. This killer probably hunted in packs, they say.

The fearsome tyrannosaur, which lived some 60 million years later than the newly discovered utahraptor, is traditionally considered the most terrible of all land predators. But the tyrannosaur was probably much slower and less nimble than the slasher, its discoverers say, and *Tyrannosaurus rex* lacked the upright slashing claw that must have made utahraptor the terror of the early Cretaceous period, about 125 million years ago.

The first claw bone of a utahraptor was unearthed in October 1991 in a quarry in eastern Utah by Carl Limoni, a laboratory staff member of the College of Eastern Utah. After the bones of several individuals were discovered, Dr. James I. Kirkland of the Dinamation International Society identified the dinosaur as belonging to the dromaeosaurid family of dinosaurs, which includes the much smaller deinonychus (a name meaning "terrible claw"), an agile, speedy creature about the size of a wolf.

—Malcolm W. Browne, July 1992

years ago divided into northern and southern land masses, Laurasia and Gondwanaland, respectively, about 150 million years ago, toward the end of the Jurassic period.

A further breakup occurred in the Cretaceous period, as today's continents took shape and drifted their separate ways 100 million years ago. Until then, the various kinds of dinosaurs had been a global phenomenon, with species in one part of the world showing remarkably close kinship with those elsewhere. But the drifting continents ended this faunal exchange, isolating dinosaurs and leading to their evolution into increasingly divergent forms before their extinction 65 million years ago.

The new discoveries, Dr. Sereno said, should give scientists a "new understanding of how dinosaurs came to be divided geographically and how that affected their evolution in the late Cretaceous."

Dr. Sereno and an international team of colleagues reported the findings in the journal *Science*. He also described them and displayed casts of the fossils at a news conference in Washington at the National Geographic Society, a sponsor of the expedition.

In an accompanying article in *Science,* Dr. Phillip Currie, a dinosaur specialist at the Royal Tyrrell Museum in Drumheller, Alberta, praised the discoveries and said they "are changing our rapidly evolving concepts of paleogeography during the Cretaceous."

Dr. Mark A. Norell, a paleontologist at the American Museum of Natural History in New York City, said that the fossils were important because so little had been known of African dinosaurs in the Cretaceous period and that they showed that large predatory dinosaurs lived on all the continents in that period. They had evolved in different ways, but they were superficially and functionally very much like one another and probably stemmed from distant common Jurassic ancestors.

"Now we need to know more about the relationships of these predatory animals," Dr. Norell said in an interview.

The new discoveries were made in the Kem Kem region of Morocco, a hot, dry land of red sandstone near the border with Algeria and in sight of the Atlas Mountains. Dr. Sereno, 38, had already established himself as one of the most successful dinosaur hunters in the field, having made sensational discoveries in Argentina, China and elsewhere in Africa. He had determined that this region of the Sahara was a vast floodplain with rivers edged by coniferous trees in the time of the dinosaurs, and he hoped that it would be rich in fossils from that period.

Dr. Sereno found the skull and five-inch-long teeth of *Carcharodontosaurus* embedded in a sandstone cliff. After further analysis, the team identified the species and recognized that it bore resemblances to *Acrocanthosaurus,* a huge meat-eating allosauroid that lived in North America during the early Cretaceous period. This indicated that some connecting land bridges must have still existed between the northern and southern land masses at that time.

The size of the new specimen led Dr. Sereno and his colleagues to believe the African predator had dethroned *Tyrannosaurus* as the king of the predators. But a few weeks later, in September 1995, paleontologists announced the discovery of what may be an even larger predator, *Giganotosaurus carolinni,* in Argentina.

It was "amazing," Dr. Sereno said, to see that predators in such widespread lands should independently reach such huge sizes.

Gabrielle Lyon, a team member who once studied at the University of Chicago, literally stumbled on the bones of what proved to be the *Deltadromeus* skeleton. A close examination revealed that this animal bore similarities to coelurosaurs, which ranged from bird-like creatures to *Tyrannosaurus*. *Deltadromeus* most closely resembled the smaller *Ornitholestes,* an agile, six-foot-long predator found in the northern continents.

Dr. Phillip Currie of the Tyrrell Museum said the discoveries showed clearly that many families of dinosaurs from around the world "were free to intermix well into the Cretaceous."

—JOHN NOBLE WILFORD, May 1996

Computer Recreates Call of Dinosaur Sound Organ

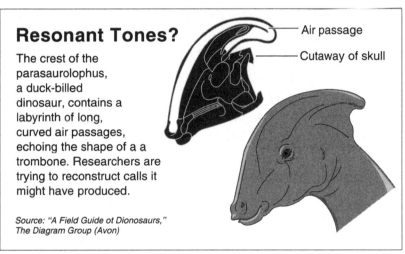

Resonant Tones?

The crest of the parasaurolophus, a duck-billed dinosaur, contains a labyrinth of long, curved air passages, echoing the shape of a a trombone. Researchers are trying to reconstruct calls it might have produced.

Air passage

Cutaway of skull

Source: "A Field Guide ot Dionosaurs," The Diagram Group (Avon)

The New York Times

MOST OF THE SCREECHES, honks and roars emitted by movie dinosaurs are based on little more than imagination, and paleontologists generally regard these cinematic noises as art rather than scientific replicas of the originals.

But with the help of powerful computers sometimes used to design nuclear weapons, scientists analyzing a recently discovered dinosaur skull hope to create a sound that may be fairly close to the call of the living animal.

The project is led by Dr. Carl Diegert, Dr. George Davidson and Dr. Constantine Pavlakos at Sandia National Laboratory in Albuquerque, New Mexico. Sandia, which is administered by the Department of Energy, has long been a major contributor to weapons research, including nuclear weapons. It is therefore equipped with some of the fastest, most powerful computers in the world.

37

The laboratory's study of dinosaur vocalization began after the discovery in August 1995 of a rare parasaurolophus (pronounced para-sore-AHLO-foos) fossil in northwestern New Mexico. The discoverer, Dr. Robert M. Sullivan, a curator at the State Museum of Pennsylvania, and Dr. Thomas E. Williamson, a curator at the New Mexico Museum of Natural History, have collaborated with the Sandia computer experts since then.

Parasaurolophuses were large plant-eating dinosaurs belonging to a numerous clan known as hadrosaurs, or "duckbills." Unlike other members of the group, the parasaurolophus sported a bony tubular crest extending back some four and a half feet from the top of its head that has long aroused the curiosity of scientists. No one can be certain of the purpose of this crest, but because it contains a labyrinth of air cavities and is shaped something like a trombone, some paleontologists surmise that it might have helped the animal produce distinctive calls, perhaps to attract mates, warn its kin of predators or socialize in other ways.

In 1981, Dr. David B. Weishampel, who is a professor of anatomy and a specialist in dinosaur studies at Johns Hopkins University School of Medicine in Baltimore, undertook a similar study of the parasaurolophus crest, eventually building a replica of polyvinyl chloride tubing and fitting it with a trumpet mouthpiece. He concluded in a paper that became his master's thesis that the crest was an excellent resonator that could emit powerful, low-pitched sounds, perhaps comparable with the ultralow notes by which elephants communicate.

"The parasaurolophus crest is highly modified with connections to the nasal cavity, the back of the throat and the lungs," Dr. Weishampel said in a recent interview.

But many questions about the animal's repertoire of sounds remained unanswered, and the Sandia group is addressing some of them with the help of computer reconstructions of the original shape of the fossil skull and computer analyses of resulting effects on the pitch, overtones, muting and other details of the dinosaur's sounds.

"The basic frequency of the note is set by the length of the tube," Dr. Diegert said, "but we're interested in much more than that. We're trying to refine the work done in the 1980's by Dr. Weishampel. For instance, we want to know not only what the sustained sound of the animal might have been, but also what its acoustic attack might have been like—the start of the animal's call, comparable to the attack of a trombonist as he begins to play."

Dr. Diegert said only four fossils of parasaurolophuses exist, and all of them were deformed by compression within the rock strata where they lay buried for about 70 million years. The specimen recovered last August must therefore be restored to its original shape and contours before detailed analysis of its acoustics will be possible.

The restoration does not involve gluing bits of bone together, however. X-ray scans of the fossil bones are being assembled as a computer image, taking into account the deformation the skull underwent, as well as the symmetry of the skull and probable locations of each piece.

"For some of this work we're using computer programs developed for the movie industry," Dr. Diegert said. "The computers we use make analyses possible that were not possible in the past. They allow you to ask, is this way of putting pieces together consistent with the data you have? You can test high-level assumptions and ask whether a part of a skull probably had one chamber or two, for example."

When the Sandia team completes its computer reconstruction of the parasaurolophus skull, it will test the fossil's acoustic properties by computer analysis and with the help of a new technique called stereolithography, in which three-dimensional laser images of objects can be used to form three-dimensional plastic replicas.

—MALCOLM W. BROWNE, March 1996

New Dinosaur Exhibit Underscores Disputes Within Paleontology

DINOSAURS were warm-blooded, colorfully marked, gregarious creatures that made terrifying noises but nurtured their young and finally died out when the Earth was hit by a comet. Or so some say.

Dinosaurian speculations like these have so rancorously divided experts in recent decades that the profession of vertebrate paleontology has sometimes seemed more like a debating society than a branch of science. Some experts believe inferences can usefully be drawn from trackways, nests and other evidence about the appearance and behavior of dinosaurs. But their opponents argue that speculation about such matters as the skin coloration of dinosaurs and their habits can never be proved or disproved, and that science should content itself with demonstrable facts unembellished by intriguing but intrinsically untestable speculations.

Unfortunately, lay dinosaur lovers rarely get to examine the arcane arguments advanced by partisans of these opposing camps, since museums traditionally adorn their fossil displays with simple labels containing little science and few references to the erudite rows that constantly rack paleontology.

An exception is the renovated exhibition at the American Museum of Natural History in New York City that outlines some of the major current debates in paleontology, acknowledging that science cannot honestly settle many of the popular speculations about dinosaurs. So even-handed does the exhibition attempt to be that many explanatory signs are adorned with caveats designed to look like the Surgeon General's warning on cigarette packs. The visitor is advised by these little green signs to be cautious in assessing all scientific debates in which the scientists themselves lack enough information to be sure.

But the very caveats offered by the museum are contentious and seem certain to arouse strong objections from some scientists.

Since the beginning of this century, the museum has acquired, mounted and displayed the biggest and best dinosaur collection in the world. The two dinosaur halls on the museum's fourth floor have always featured the great fossils themselves, with simple written explanations of relationships. But with the four-year, $12 million renovation, visitors are treated not only to new and thoroughly refurbished incarnations of the familiar fossils (as well as some new acquisitions) but to a generous helping of the science inspired by the bones of the great beasts.

American Museum scientists who supervised the renovation, including Dr. Lowell Dingus, Dr. Eugene S. Gaffney, Dr. Mark A. Norell and Dr. Michael J. Novacek, the museum's provost of science, were at pains to correct some conspicuous errors of previous displays for which the museum came in for criticism.

For example, it had been known since the 1970's that the museum's first big dinosaur fossil to go on view, a gigantic *Apatosaurus* first displayed in 1905, had the wrong head. Early paleontologists mistook the skull of a big dinosaur named *Camarasaurus* for that of *Apatosaurus*, and for more than 80 years, the museum's star dinosaur faced visitors with a small, blunt skull from a different genus.

At the time, museum officials decided to leave the wrong head in place until a complete renovation of the fossil was possible, since a simple replacement of the skull could have triggered a catastrophic collapse of parts of the skeleton. The fossil now has the right head, a more accurate stance, several new cervical vertebrae that had been overlooked originally and almost 20 more feet of tail assembled from replicas of tail vertebrae now believed to have belonged to the animal. (The fossil is now 86 feet long, and its former head is now displayed in a case under it.)

Another of the museum's most famous fossils, *Tyrannosaurus rex*, also embodied some errors that have been corrected for the new display. Paleontologists formerly believed that tyrannosaurs walked with an upright stance, supported in part by their powerful tails, which supposedly dragged along the ground. But because the fossilized footprints of dinosaurs rarely show any evidence that their tails touched the ground, and because paleontologists discovered by calculations that upright posture would have placed too great a load on part of the tyrannosaur's spinal column, the big fossil had to be completely remounted.

In its present guise, the 3,500-pound fossil is mounted with its head low and its body parallel to the ground, pivoting on its huge pelvis with the tail as a counterweight to its body. Its neck has been reconfigured as slightly S-shaped, in accord with the latest thinking. To pose the big dinosaur in a realistic hunting stance, the museum first raised the fossil on large jacks and then suspended parts of it from thin, strong cables attached to concealed anchor beams in the ceiling.

Heavy manual labor went into the fossil restoration. Jeanne Kelly, the museum's chief preparator, said that it took six people working for two months with alcohol solvent just to strip away a century-old layer of shellac that had covered (and discolored) the tyrannosaur's mineralized bones.

Phillip Frailey, a museum expert in mounting technology, supervised the casting of heatproof cement replicas of the nearly 200 bones in the tyrannosaur—77 vertebrae and 24 ribs among them. His team used the cement replicas as templates for casting a strong but unobtrusive steel armature, or support structure. Each of the bones in the finished display now fits into this scarcely visible steel framework in modular fashion, so that in the future, an individual bone can be removed for study by scientists without disturbing the rest of the huge skeleton.

"In our old exhibit," Dr. Norell said, "the *Tyrannosaurus* looked like Godzilla. Now it looks like a bird."

Scientists generally now agree about how tyrannosaurs and their kin walked in life. But much less is known about many other extinct species dating from the time of the dinosaurs, and fragmentary knowledge is complicating the work of museum preparators.

"At present we're working on a fossil of the marine plesiosaur *Elasmosaurus*," Mr. Frailey said, "and we do not know its mode of locomotion. With its very long neck it might have propelled itself by wriggling like a snake, but it also had fins that might have been its prime movers."

Despite the museum's effort to avoid seeming to take sides in scientific disputes, the new exhibition's displays and explanatory notes are certain to provoke criticism from some paleontologists.

For example, the exhibition designers accepted as proved the hypothesis that one branch of the dinosaur group, the birds, survive to this day. Entering the new Hall of Saurischian Dinosaurs a visitor walks under a suspended flock of modern gulls represented as close relatives of the tyrannosaur looming behind them. Labels, explanations and accompanying pamphlets invariably refer to extinct dinosaurs as "non-avian dinosaurs," meaning that "avian dinosaurs"—birds—are not extinct.

Dr. Norell and other Museum of Natural History scientists base their belief in the kinship of birds and dinosaurs on shared "characters"—distinctive physical features—and on a system of classification called cladistics, which was pioneered at the museum. Cladistic analysis, unlike earlier systems for comparing species, does not attempt to show the chronology of evolutionary change, but merely places species along the branches of a diagrammatic tree in which the distance between any two species is defined by the number of their shared characters.

Cladism is widely accepted as a useful way to show relationships, but it is still challenged by some paleontologists on the grounds that by ignoring chronology it is sometimes subject to serious errors. In Britain during the 1970's and 1980's, arguments over the merits of cladism deeply embittered relations among scientists at the British Museum (Natural History). Some argued that cladism is too easily exploited by religious creationists as a way to explain the diversity of life without having to invoke evolutionary theory, which is opposed by the creationists.

The new exhibit directly challenges some widely held hypotheses, including the belief by some prominent paleontologists that certain species of dinosaurs nurtured their young in protective nests and communities long after hatching. Museum explanations do not mention by name John Horner, a paleontologist at the University of Montana in Bozeman, but some comments are clearly aimed at him and his work.

One display briefly summarizes Mr. Horner's famous discovery of preserved hadrosaur babies, adolescents and eggshells buried in volcanic ash in what appear to have been carefully designed nests. (Mr. Horner named the dinosaur genus *Maiasaura*, or "fostering lizard.") But the museum placard says that modern crocodiles leave similar nests containing both hatchlings and adults, even though crocodiles guard their young for only very brief periods. A green "warning" label adds: "Consequently, this evidence does not allow us to draw conclusions about parental care in extinct dinosaurs. We can only make guesses based on what we know of their living relatives."

Designers of the new exhibition treated the great controversy over the cause of "non-avian dinosaur" extinction in much the same way; a sign informs visitors that at the time dinosaurs died out 65 million years ago, there is evidence of gigantic volcanic eruptions as well as evidence that large objects from space hit the Earth.

Either volcanism or a giant meteorite could have killed off the dinosaurs (other than the dinosaurs that had evolved as birds), the museum says, but the fossil record is too spotty to reveal what really happened.

Such institutional assertions by the museum make some paleontologists bristle.

—MALCOLM W. BROWNE, May 1995

Familiar Dinosaurs May Take New Shape

Stephen A. Czerkas

NEW FOSSIL DISCOVERIES in Wyoming show that paleontologists, and all those vivid images in museums and books, could be wrong in one very visible respect about the appearance of the hulking sauropods, one of the most familiar groups of dinosaurs.

The long tails of these huge reptiles, it now seems, were not smooth and unadorned. In at least some fossil specimens, a row of spines ran along the top ridge of the tail, raising speculation that the spines continued along the top of the body and neck as well.

In reporting these surprising findings at the annual meeting of the Society of Vertebrate Paleontology in Toronto, Stephen A. Czerkas, an independent paleontologist and dinosaur illustrator, said this would "open the floodgates for reinterpretation" of sauropods, including species like *Apatosaurus* (commonly called *Brontosaurus*), *Barosaurus* and *Diplodocus*. This could result in major revisions of these animals as represented in popular exhibits and scientific literature and drawn with fond expertise by schoolchildren.

In recent decades, there has been an increasingly mammalian connotation in portrayals of the largest sauropods, with their long, slender necks and tails extending from elephantine bodies with smooth, leathery skin. This could have influenced, or been influenced by, research suggesting that dinosaurs may have been more like warm-blooded creatures, in the manner of mammals and birds, than the usual cold-blooded reptiles.

"Now we're seeing that at least some sauropods really looked reptilian, as people think of reptiles, with spikes that remind you of very large iguanas," Mr. Czerkas said in an interview. "This could shift some scientific thinking."

Mr. Czerkas, who is co-director of a dinosaur museum being established in Blanding, Utah, reported the discovery at the meeting in Toronto. A more detailed analysis of the research, along with recreations of what the spiked sauropods might have looked like, will be published in the journal *Geology.*

Paleontologists were cautious in their initial reactions to the research. They were no doubt mindful of mistaken inferences in the past. For years, museums placed the wrong skull atop their brontosaur skeletons. In the 19th century, one of the earliest dinosaur experts, Gideon Algernon Mantell, mistook the spiky thumb bone of the *Iguanodon* for a horn and, in drawings, placed it on top of the snout, making the creature look like a rhinoceros.

But paleontologists praised Mr. Czerkas as a careful interpreter of how dinosaurs looked. They also noted that a few previous fossil hunters had found remains similar to the tail spikes cited by Mr. Czerkas, but had never considered their possible significance.

"If we find more evidence like this, we'll be much happier," said Dr. Kevin Padian, a paleontologist at the University of California at Berkeley, who has examined the fossil material. "Then we may be convinced there is no other reasonable explanation."

The evidence for the tail spikes was uncovered at Howe Quarry, a site near the town of Greybull in northern Wyoming that had produced many dinosaur discoveries earlier this century. New excavations were started in 1990 by Kirby Siber, a commercial collector from Switzerland. When he came upon fossil skin impressions in the stone, he called in Mr. Czerkas for a look. A few years ago, his interpretation of new discoveries had revealed that the characteristic bony plates on the backs of stegosaurs were arranged in a single row, not in a double row.

In the article for *Geology,* Mr. Czerkas said many conical, spine-like fragments were found in the quarry, some isolated but others connected together, "indicating that their natural positions in life formed a continuous line." In two cases, the fossils showed that the spines were "located along the dorsal midline structure, at least along the whiplash portion of the tail."

Mr. Czerkas said, "Exactly how far the spines continued up the tail is not known. And the precise pattern and full extent of the ornamentation also remains unknown. But, as with the hadrosaurs, it is likely that the spines continued beyond the tail and along the sauropods' body and neck as well."

Paleontologists have not identified the species from which the skin impressions came. They said the fossils resembled dinosaurs of the diplodocid family, a group of the plant-eating sauropod giants that includes *Diplodocus* and *Barosaurus.* More than a dozen examples of spikes were clearly identified, ranging in size from two inches upward of nine inches. The smaller ones were at the very end of the tail. Some were sharply pointed and straight; others were curved and may have had blunt tips.

Unlike armored dinosaurs like the stegosaurs or ankylosaurs, these sauropod skeletons show no obvious signs in the spine that the animals had any exterior ornamentation like spikes or plates. From the impressions, Mr. Czerkas said, it was impossible to tell what the spikes were made of, but it was not bone. Dr. Padian said the most likely composition was keratin, the protein in fingernails, claws and hair.

Mr. Czerkas said the discovery at Howe Quarry "implies, but does not prove, that other types of sauropods were similarly equipped." But he suggested that museums and illustrators would soon be forced to modify their images of these popular creatures.

"Until additional physical evidence demonstrates otherwise," Mr. Czerkas said in the *Geology* article, "the traditional imagery of sauropods without dermal spines is contrary to the evidence at hand and the current understanding of what these dinosaurs actually looked like."

—JOHN NOBLE WILFORD, November 1992

Chemical Traces of Blood Found in Bones of *Tyrannosaurus Rex*

SCIENTISTS SAY they have recovered elements of blood from the bones of a dinosaur that died some 65 million years ago. This is the first time that blood components have been recovered from dinosaur bones, and the nearest that science has come to the science fiction fantasy of recreating the creatures from the genetic information in their blood cells.

The dinosaur, a nearly complete *Tyrannosaurus rex*, was buried in conditions that prevented its bones from being converted to mineral, as is the case with most fossils. The interior of the bones are preserved largely in their original form.

Dr. Mary H. Schweitzer, a paleontologist at Montana State University in Bozeman, recovered the specimen from the Hells Creek formation of eastern Montana. Because of the bones' state of preservation, she decided to look for direct signs of life, like cells, DNA and protein.

The recovery of ancient molecules is fascinating but treacherous terrain on which many scientists have stumbled. The more sensitive the analytical technique, the higher the risk of picking up molecules that have contaminated the original specimen. The DNA that a team of Utah scientists recently asserted that they had recovered from 80-million-year-old dinosaur bones was later said by other researchers to be of human origin.

Because of such setbacks, Dr. Schweitzer and her colleagues proceeded with caution and took several years to publish their results.

Earlier hopes of finding cells in the dinosaur bone have been dashed. Dr. Schweitzer said she could see no direct sign of cells, although a chemical stain that recognizes DNA picked up something in the holes where the bone cells would have rested.

But she said she had been unable to retrieve DNA that could be identified as originating in a dinosaur.

She and her colleagues had better luck in looking for heme, the oxygen-carrying part of the hemoglobin molecule of the blood.

They report finding evidence for the presence of heme by six different chemical tests in an article published in *The Proceedings of the National Academy of Sciences.*

In one of the tests, they injected ground-up dinosaur bone into rats to see if the animals would become immunized against dinosaur blood. Since no dinosaur was at hand to check if the vaccination was successful, a turkey was chosen as a stand-in, on the theory that if birds descended from dinosaurs the turkey could be a close-enough cousin to *Tyrannosaurus rex,* at least immunologically speaking. The rats did react to turkey blood, evidence that they had made antibodies to some part of the hemoglobin molecule.

Dr. Schweitzer said the rats' immune systems were probably reacting to a fragment of hemoglobin, not the whole molecule. Another chemical test suggested that the heme was attached to a protein fragment.

Dr. S. Blair Hedges, an evolutionary biologist at Pennsylvania State University, said he had found Dr. Schweitzer's article convincing, since she had gathered "a pretty substantial body of evidence supporting the preservation of these biomolecules in dinosaur bone."

Dr. Norman R. Pace, the member of the National Academy of Sciences who recommended Dr. Schweitzer's article for publication in the academy's journal, said, "I think she did a real careful job, the best that can be done with present technology."

The significance of Dr. Schweitzer's work probably lies in its message to other paleontologists that biological molecules are worth looking for in fossils, though carefully.

The long-standing belief that proteins and DNA could not survive for millions of years has not been greatly changed by recent assertions to the contrary, since many of these claims are widely doubted.

Dr. Schweitzer said that the lightly mineralized state of the dinosaur was unusual but far from unique and that she believed that molecular evidence might be preserved in fossils more commonly than thought. That feeling is apparently shared by Dr. Pace. Calling Dr. Schweitzer's report "a bit of a bombshell," he said he had recommended it for publication because "the field needed stirring up."

Genuine DNA data from fossils would be greatly prized because its genetic information would help reconstruct the pathways of evolution. Protein data, Dr. Schweitzer suggested, could yield immunological information that would clarify the relationships among extinct species.

There have been sporadic assertions in the past of finding very ancient protein fragments in fossils, and more frequent claims of ancient DNA from specimens preserved in amber. But few of these can be repeated, and many researchers who have looked for ancient DNA in amber have failed to find any. Since DNA has limited chemical stability and is kept in repair in living organisms by an elaborate suite of enzymes, it seems likely in normal circumstances to decay rapidly after the death of the cells that maintain it.

—Nicholas Wade, June 1997

Pity a Tyrannosaur? Sue Had Gout

FOR ALL THE SUFFERING she probably caused her Cretaceous prey, a tyrannosaur named Sue seems to have paid dearly. Scientists have determined that the big dinosaur probably was a victim of agonizing gout and other debilitating ailments.

The diagnosis of dinosaurian gout, published in the journal *Nature,* was made by Dr. Bruce M. Rothschild of the Arthritis Center of Northeast Ohio in Youngstown, and two paleontologists, Dr. Ken Carpenter of the Denver Museum of Natural History and Dr. Darren Tanke of the Royal Tyrrell Museum in Drumheller, Alberta.

The fossil at the focus of their inquiry, called Sue by its discoverers (although the sex of the animal is uncertain), is one of the most complete tyrannosaur skeletons ever found. But the splendid specimen has brought little but misfortune to its discoverers. Now, it appears, Sue herself was unlucky.

The scientists who examined some of the dinosaur's bones have concluded that the equivalent of hand bones (metacarpals) at the ends of the animal's spindly forearms are riddled with lesions strikingly similar to those observed in some modern animals—birds, reptiles and humans, for example—with gout. Gout is a metabolic disease in which uric acid erodes bone tissues and deposits sharp mineral crystals in joints, causing excruciating pain.

Earlier studies had found that Sue's facial bones had been slashed by deep gashes, a dinosaur tooth was left embedded in one of her ribs and she probably limped from a badly healed broken leg.

"Sue was not a well dinosaur," Dr. Carpenter said in an interview. "If she was as bad-tempered as we're inclined to think, she certainly had her reasons." Sue was found in 1990 protruding from a South Dakota butte by a team of commercial fossil dealers headed by Peter L. Larson of Hill City,

South Dakota. Mr. Larson paid Maurice Williams, the Indian on whose land the fossil was found, $5,000 for the right to excavate and remove Sue.

But in 1992, FBI agents and National Guard troops under orders from the Justice Department raided Mr. Larson's workshop and seized Sue and many other specimens, on the grounds that they had been illegally removed from government land, an Indian reservation.

Mr. Larson was subsequently convicted of carrying currency illegally between the United States and two foreign countries in connection with dealing in fossils. He was sentenced to two years in Federal prison and is scheduled for release in August.

The ownership of Sue, meanwhile, has been assigned by the courts and various Federal agencies to Mr. Williams. Sue is now scheduled for auction on Mr. Williams's behalf by Sotheby's this fall.

When the Federal Bureau of Investigation seized and impounded Sue, the fossil was mostly encased in the rock in which it had lain for some 65 million years. But before the seizure, Mr. Larson had his colleagues had freed the skull and some body parts from the rock, and Mr. Larson invited Dr. Carpenter to examine the exposed fossil bone and make a cast of one of the forelimbs.

Later, at a meeting in Denver, Dr. Carpenter showed the cast to Dr. Rothschild, a specialist in bone disease, who immediately noticed lesions in the hand bones characteristic of gout.

The team's subsequent examination of the forearm, which included X rays and other advanced techniques, failed to find any crystals of gouty uric acid or urates, which would have been dissolved away soon after the animal's death. But they did find various bony malformations, in which bone had been eaten away by some process most likely to have been gout.

"There are one or two extremely rare pathologies that might be responsible for lesions like these," Dr. Rothschild said in an interview, "but gout is by far the most likely explanation." Tyrannosaurs were meat eaters, and popular opinion used to attribute gout to diets that included too much red meat and wine.

"The wine-drinking part had some basis," Dr. Rothschild said, "because some wines were bottled in containers of which lead was an ingredient. Lead poisoning can cause gout lesions. But in Sue's case I think we can rule it out." Pre-disposition to gout is now regarded as genetic. Gout occurs in animals that do not properly metabolize purines (some building blocks of DNA) and therefore pour excessive uric acid into the bloodstream.

Dr. Carpenter said the group examined only arm and hand bones and was unable to look for evidence of gout in Sue's feet before the fossil was seized.

But to broaden their study of dinosaur gout, Dr. Rothschild and Dr. Carpenter enlisted the collaboration of Dr. Tanke of the Royal Tyrrell Museum, which owns an enormous collection of dinosaur fossils. The group examined 83 finger bones of tyrannosaurids (various species belonging to the genus *Tyrannosaurus*) and found only one bone exhibiting evidence of gout.

Their conclusion: Gout was probably rare among tyrannosaurs, and Sue may just have been born under an evil star.

—MALCOLM W. BROWNE, May 1997

Embryo Sheds Light
on Dinosaurs' Origin

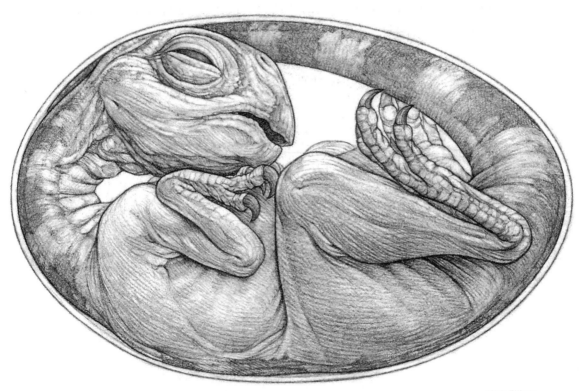

Mick Ellison

IN A DISCOVERY that should begin to round out knowledge of the full life cycle of dinosaurs, paleontologists exploring the Gobi Desert of Mongolia have found the first fossilized embryo of a meat-eating dinosaur. Only six or seven other dinosaur embryos are known to science, and none is as exquisitely preserved as this one.

54

Curled up and resting in part of its oblong egg, the specimen looks very much like a tiny dinosaur on the half shell. With tail and all, the fully extended embryo would probably measure eight inches long. But it is still in the fetal position, the head tucked near the knees. A hand is over the face. Except for the missing pieces of the tail and the top of the skull, everything about the skeleton seems complete, with individual vertebrae, pelvic bones and limbs all well formed and clearly identifiable.

In his laboratory this week, Dr. Mark A. Norell of the American Museum of Natural History in New York, who made the discovery, tenderly cupped the 75-million-year-old specimen in his hand. It was strange to see a hunter of the ancient reptiles best known for their fierce teeth, sharp claws and gargantuan bones smiling over a find so delicate and vulnerable.

"I knew this was an embryo as soon as I saw it lying on the ground," Dr. Norell said. "I knew from the ankle bones it was a theropod," the broad group of dinosaurs that includes agile carnivores like *Tyrannosaurus rex,* velociraptors and the smaller, bird-like oviraptors.

An analysis in the laboratory, after the sand and minerals were painstakingly cleaned away, confirmed Dr. Norell's first impressions. It also determined that the tiny bones had already ossified, indicating that the embryo had been close to hatching.

The cranial features, Dr. Norell said, identified it as a member of the oviraptorid family, or dinosaurs that grew to be more than six feet long with short heads, elongated necks, toothless jaws and horn-like bumps on the end of their snouts. They probably looked something like an ostrich with a tail, running about on two legs and attacking prey with strong claws on their forelimbs.

Details of these findings are reported in the journal *Science* by Dr. Norell and a team of American and Mongolian scientists. The discovery was made in the western Gobi in the summer of 1993 on an expedition from the American Museum and the Mongolian Academy of Sciences, led by Dr. Michael J. Novacek, a paleontologist and provost of science at the museum.

Other scientists examined the specimen last week at a meeting of the Society of Vertebrate Paleontology in Seattle and agreed that it was an important discovery that should open a window on the early life of dinosaurs and their nesting behavior. It could also provide further support for an ancestral connection between dinosaurs and birds, as many scientists have theorized.

"This is absolutely non-controversial," Dr. Norell said in an interview. "Everyone agrees it's an oviraptorid."

John R. Horner, a dinosaur paleontologist at the Museum of the Rockies in Bozeman, Montana, praised the discovery because good dinosaur embryos are so rare. "Mostly we have adults to work with," he said, "and that only tells you about adults."

In particular, Mr. Horner said, the bone development of embryos could provide evidence of whether, as hatchlings, these dinosaurs were able to walk or had to stay in the nest for some time. An examination of the few embryos of plant-eating dinosaurs, he said, had produced conflicting results, indicating some probably could walk immediately, but others could not. Since most dinosaur eggs are rock-solid fossils, any embryos preserved inside can only be studied indirectly, and not too well, with X rays and CAT scanners.

In his excavations in Montana, Mr. Horner has made many important discoveries about the parental practices of plant-eating dinosaurs, finding eggs, shells and juvenile skeletons in large nests arranged somewhat like those in a teeming penguin colony. The remains of an adult associated with some nests suggested that parental nurturing might be part of the behavior of some dinosaurs. One of these was a duck-billed animal given the name *Maiasaura*, the good mother reptile.

The oviraptorid embryo was found at Ukhaa Tolgod, a mile-wide basin in the western Gobi that has proved to be one of the richest lodes of vertebrate fossils from the end of the age of dinosaurs. The American-Mongolian expedition discovered the site in 1993 and returned last summer for further exploration.

Erosion had exposed a clutch of fossil eggs, which were about five inches long and two and a half inches wide. One was so badly weathered that the top had worn away, revealing the tiny embryo skeleton. Besides giving scientists a rare view of early dinosaur life, the discovery immediately confronted them with a mystery of mistaken identity.

In 1923, an American Museum expedition led by Roy Chapman Andrews discovered the first known cluster of dinosaur eggs at a spectacular site in the Gobi called Flaming Cliffs. The eggs were assumed to belong to a species of plant-eating dinosaur known as *Protoceratops*, because it was the most common dinosaur fossil the explorers had found in the Gobi. They read even more into this pre-historic scene.

Lying atop the nest was the strange-looking skeleton of an unknown dinosaur. It was identified as a carnivore that probably died in a sandstorm while sucking the *Protoceratops* eggs. So the fossil was named *Oviraptor,* which means "egg seizer" in Latin, and ever since its reputation has suffered accordingly.

Dr. Norell's discovery in the basin 200 miles from the Flaming Cliffs has revealed that the dinosaur had been misnamed. Determining that these were the eggs of the supposed predator itself, not a *Protoceratops,* amounted to a vindication for *Oviraptor.* "Rather than eating the eggs, they were incubating them or protecting them," Dr. Norell said.

In the same nest the scientists uncovered two tiny skulls of another type of carnivorous dinosaur from the group known as dromaeosaurs, possibly *Velociraptor.* The skulls may have been those of embryos or newborns. But to find these skulls in the same nest with an embryo of another species is extraordinary, paleontologists said.

The little dromaeosaurids were most likely brought to the nest as food by the adult oviraptorids. Or else they may have been predators, raiding the nest, or nest parasites, as cuckoos are today. Adult cuckoos lay their eggs in the nests of birds of other species, abandoning them to be hatched and raised by the surrogate parents.

Dr. John H. Ostrom, a paleontologist who recently retired from the Yale University faculty, said the findings bolstered the case for a dinosaur-bird ancestral link. This is the first strong evidence for bird-like nesting behavior by theropods, the group of dinosaurs from which birds perhaps evolved more than 150 million years ago. Of course, these oviraptorids, having lived much later, would not have been among the direct ancestors of birds.

The other authors of the journal report of the embryo discovery, besides Dr. Norell and Dr. Novacek, were Dr. Luis M. Chiappe, Amy R. Davidson and Dr. Malcolm C. McKenna of the American Museum; Dr. James M. Clark of George Washington University; Dr. Demberlyin Dashzeveg and Dr. Barsbold Rhinchen of the Mongolian Academy, and Dr. Perle Altangerel of the Mongolian Museum of Natural History.

—JOHN NOBLE WILFORD, November 1994

Fossil of Nesting Dinosaur Strengthens Link to Modern Birds

THE REMAINS of an *Oviraptor* that died 80 million years ago in the Gobi Desert of Mongolia while sitting on its nest of eggs is the first definitive evidence of parental care among at least some dinosaurs.

Its legs were tucked tightly behind its body in a manner identical to that of such modern birds as chickens and pigeons. The arms were held back and somewhat to the side, encircling the nest in a protective posture.

Mick Ellison

Evidence of Parental Care Among Dinosaurs
Fossil hunters in the Gobi Desert have found the remains of a dinosaur that died while on its nest of eggs.

Directly under the squatting reptile were the eggs, measuring seven inches by two and a half inches and arranged in a circular pattern corresponding to the nests of many modern birds.

Something dire happened. Either the *Oviraptor* died shortly before a sandstorm struck and was buried almost immediately, or it died as a result of the storm. In any event, a quick burial deep in dry sand fossilized the animal's bones and eggs and preserved them exactly in place, a moment in time frozen for all time.

The revealing fossils were uncovered by scientists from the American Museum of Natural History in New York City and the Mongolian Academy of Sciences, who have been exploring the paleontological treasures of the Gobi each year since 1990. The discovery was made in the Ukhaa Tolgod Basin at the foot of the Nemegt Mountains, a site in the central desert that has turned out to be one of the world's richest lodes of reptile and mammal fossils from the end of the age of dinosaurs.

The *Oviraptor* fossils were excavated two years ago, but it was many months after the scientists got the chunk of red sandstone containing the fossils back to the American Museum until they realized the import of their discovery. Cautiously, they chipped at the surrounding rock, exposing the bones and the eggs.

"You could just look at it, and it jumps out at you as an animal sitting on a nest," Dr. Mark A. Norell, a paleontologist at the American Museum, said in an interview. "It was one of the most spectacular things we have ever found."

Reporting the discovery in the journal *Nature*, the team of scientists led by Dr. Norell concluded, "This finding provides the strongest evidence yet that modern avian brooding behavior evolved long before the origin of modern birds," and occurred in some dinosaurs.

Dr. David B. Weishampel, a specialist in dinosaur anatomy at Johns Hopkins University in Baltimore, agreed that the find provided for the first time "astonishing and incontrovertible evidence" that some dinosaurs had nesting behaviors almost identical to those of birds. The fossils, Dr. Weishampel said, open up "a wealth of research avenues on dinosaurian life histories."

Of perhaps greatest significance, scientists said, the newly revealed brooding behavior fits neatly into the emerging and increasingly compelling image of dinosaurs as close relatives of birds. Such a relationship has been considered off and on since the 19th-century discovery of *Archaeopteryx,* the earliest known bird, which resembled a small dinosaur.

"Skeletal evidence clearly indicates that birds are a kind of living dinosaur," said Dr. Luis M. Chiappe, another member of the discovery team from the American Museum. "The unique anatomical characteristics uniting these two groups have been well documented and recognized for years. This discovery proves for the first time that birds and dinosaurs also share complex behaviors."

The two other members of the expedition who were co-authors of the report are Dr. James M. Clark, a paleontologist formerly of the American Museum who is now at George Washington University, and Dr. Demberlyin Dashzeveg, a paleontologist with the Mongolian Academy of Sciences in Ulan Bator. Dr. Norell is the one who made the initial discovery in the field, seeing some fossilized claws sticking out of the red soil of the basin floor.

In their report, the scientists said the specimen was the first preserved well enough to determine its precise relationship with the nest—that is, its body was covering the eggs in the nest in "the same posture taken by many living birds when brooding." The eight-foot-long specimen is largely complete, except for the skull, pieces of a hind limb and parts of the vertebral column.

Fifteen eggs have been identified, with several more likely to be concealed beneath the top layer. The eggs are arranged in a circle, invariably with the broad end pointing toward the center of the nest.

Paleontologists still have many unanswered questions about the nesting specimen.

Was the dinosaur male or female? Among many bird species, males and females share nesting duties. Dr. Norell said scientists would examine the microstructure of long bones in the specimen to search for excess calcium deposits that might show that the dinosaur was a female. In birds, such deposits are associated with the female's egg production.

What was the purpose of the brooding behavior? In this case, paleontologists said, the dinosaur might have been either shading the eggs from the desert heat or guarding them from predators. Most modern birds incubate their eggs to maintain a warm constant temperature, but it is not known whether this was the original purpose of nesting or is a more recent behavior.

The discovery team said no other eggs had been found close enough to suggest that the specimen was part of a nesting colony. Mr. Horner found such colonies to be common at dinosaur sites he explored in Montana.

The paleontologists said they were not ready to draw from the new evidence any conclusions about the possibility that dinosaurs were a type of warm-blooded reptile—not cold-blooded, as is the rule among reptiles. But if the nesting dinosaur was incubating the eggs with body heat, they said, this would be "strongly suggestive" that dinosaurs possessed some reptilian version of warm-bloodedness.

—JOHN NOBLE WILFORD, December 1995

In China, a Spectacular Trove of Dinosaur Fossils Is Found

AN INTERNATIONAL TEAM of paleontologists has discovered a fabulous trove of dinosaur fossils in a remote region of northeast China. Hundreds of major finds at the site include the first fossilized internal organs of dinosaurs ever seen, and the first fossil of a dinosaur containing a mammal it had just eaten.

Many of the excavated specimens seem to bear on the question of kinship between dinosaurs and birds, although the wealth of enigmatic fossils seems more likely at this stage to inflame debates over the origin of birds than to settle them.

Among specimens recently recovered from the site by a Chinese team were more than 200 fossils of a primitive bird, *Confucius ornis,* together with many species of dinosaurs, mammals, insects and plants, an apparently complete record of the life there at the instant in the late Jurassic or early Cretaceous period when it was suddenly wiped out.

The scientists surmise that a brief but lethal catastrophe, perhaps a huge volcanic eruption, killed and buried everything there, possibly even bacteria.

The announcement of the find and of an agreement by China to permit international cooperation in the study of the site was made at a meeting at the Academy of Natural Sciences in Philadelphia.

Four American paleontologists and one German paleontologist, chosen because of the diversity of their experience and views, reported on a reconnaissance trip to China.

The site, near the village of Beipiao in Liaoning Province in northeastern China, was discovered by a local farmer who realized the potential scientific value of a fossil he found, which looked as much like a bird as a dinosaur, and which seemed to have a feathery crest—perhaps some form of primitive feathers or fur. He split the rocky specimen in half, selling one half to scientific institutions in Beijing and the other half to their rivals in Nanjing.

Photographs of this transitional animal were presented in October 1996 in New York City at a meeting of the Society of Vertebrate Paleontology.

The presentation caused such a sensation that Dr. Donald Wolberg of the Philadelphia academy arranged for a reconnaissance trip by international experts to inspect the site.

Dr. John H. Ostrom, a retired Yale University professor who in 1964 discovered the velociraptor dinosaurs made famous in the movie *Jurassic Park*, headed the inspection in China. "This was one of the most exciting moments of my life," Dr. Ostrom said. "I'm not aware of any richer fossil sites anywhere."

The others in the group were Dr. Peter Wellnhofen of the University of Munich, Germany, a specialist in fossil birds; David Bubier of the Philadelphia academy; Dr. Larry D. Martin of the University of Kansas, and Dr. Alan Brush of the University of Connecticut.

Among the fossils the group examined were several specimens of sinosauropteryx, an animal very similar to compsagnathus dinosaurs. In one fossil they found an oviduct containing an egg that would have been laid if the animal had lived. This was the first fossilized internal organ ever found in a dinosaur.

In another specimen of the same species they found the fossil jawbone from a primitive mammal the dinosaur had just eaten. The jawbone is about one inch wide and is studded with sharp little teeth.

The main immediate interest of the group focused on specimens that might or might not represent transitional life forms between dinosaurs and birds.

Although a majority of paleontologists accept the view that modern birds are direct descendants of dinosaurs, other experts argue that the two lines are merely "convergent"—that they evolved into similar forms through a process of parallel evolution, not by direct kinship.

Dr. Ostrom and others believe that a kind of black line running along the spine of the sinosauropteryx is a fossil remnant of something like fur or possibly feathers. He has long argued that birds, which have skeletons strikingly similar to those of theropod dinosaurs, descended directly from dinosaurs.

Others in his group disagreed, arguing that the black frill along the animal's spine does not resemble feathers but could actually be the residue of tissues that lay under the skin of the animal in life. It could also have been left by bacteria as the animal decomposed.

Another enigmatic Chinese fossil prompted debate among the team members. The chicken-size dinosaur, dubbed by Chinese scientists as protoarchaeopteryx, is believed by these scientists to be a primitive, nonfeathered relative of archaeopteryx, the earliest known true bird.

But Dr. Martin and several others challenged this on the ground that protoarchaeopteryx apparently lived later than archaeopteryx and could not therefore have been its ancestor.

The expedition to northeast China raised many more questions than it answered, and participants agreed that much more investigation is required. For one thing, the dates of the fossils are somewhat uncertain because geologists are not sure about when the sediments in which the fossils are embedded were laid down.

"I look forward to a wonderful cooperative project with American and other international paleontologists," Dr. Ji Qiang, director of the National Geological Museum of China, said. "This locality we have just begun to look at is not only a Chinese treasure, it is a global treasure."

—MALCOLM W. BROWNE, April 1997

An Early Bird Mars Theory on Dinosaurs

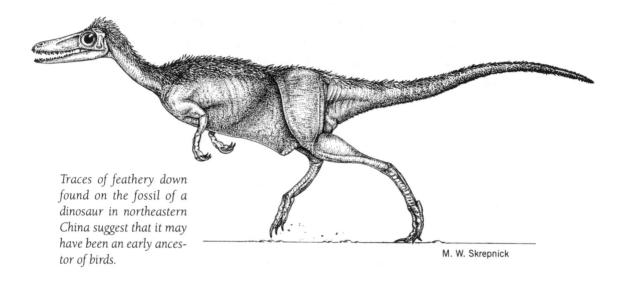

Traces of feathery down found on the fossil of a dinosaur in northeastern China suggest that it may have been an early ancestor of birds.

M. W. Skrepnick

IN AN ANALYSIS of new fossil discoveries in northeastern China, American and Chinese researchers have found evidence that they say casts serious doubt on the widely held theory that birds are direct descendants of dinosaurs.

The findings suggest that archaeopteryx, a reptilian bird that lived 150 million to 145 million years ago, and its close kin were not direct links between reptiles and today's birds, but evolutionary dead ends. By the time of archaeopteryx, another bird lineage with perhaps much more ancient origins existed. That lineage seems to have led to modern birds.

The strongest clue, scientists said, comes from the fossils of a bird the size of a sparrow, called *Liaoningornis*, which was uncovered by a farmer in Liaoning Province. In one test, volcanic rocks associated with the bones have been dated at 137 million to 142 million years old, which in evolutionary terms would make the bird a virtual contemporary of archaeopteryx.

But the specimen is considerably more modern-looking than the primitive archaeopteryx, having foot bones and a keeled breastbone that resemble those of birds today. It offers the earliest evidence of such a distinctly bird-like structure as the keeled breastbone, which is essential for flight. (It is not clear whether archaeopteryx, which did not have one, was capable of more than gliding downward from cliff tops.) The bones also suggest that this bird was probably warm-blooded, like modern birds and unlike reptiles.

In a report published in the journal *Science,* the researchers concluded that this and other discoveries indicated a much more ancient avian history, extending perhaps to reptilian ancestors even before the age of dinosaurs, which began more than 200 million years ago.

"This exacerbates one of the most obvious conundrums facing the theory of a dinosaurian origin of birds," the scientists wrote. "The dinosaurs thought to be most like birds are younger than archaeopteryx by more than 76 million years."

This time paradox, the scientists suggested, has led some to speculate that birds gave rise to certain dinosaurs in the late Cretaceous geologic period, rather than the reverse. The Cretaceous ended 65 million years ago, the time of the mass extinction that wiped out dinosaurs and most birds. Those that survived, mainly shorebirds, then evolved rapidly into the bird groups known today.

In any event, one of the scientists, Dr. Alan Feduccia, an ornithologist at the University of North Carolina at Chapel Hill, said the fossils offered "strong evidence for a fundamental split in early bird evolution" long before archaeopteryx. "This raises an interesting question about whether archaeopteryx and all of the other recently discovered early birds may be a sideline of avian history," he said in a telephone interview.

The other contributors to the *Science* report are paleontologists: Dr. Lianhai Hou of the Institute of Vertebrate Paleontology in Beijing; Dr. Larry D. Martin of the University of Kansas at Lawrence, and Zhonghe Zhou, a graduate student at Kansas.

Their interpretation of the new fossils is certain to ruffle feathers among paleontologists, most of whom embrace the theory that birds evolved from dinosaurs, though fewer and fewer still think that archaeopteryx was the direct ancestor. Ornithologists, on the other hand, have long disputed the dinosaur-bird link, contending that dinosaurs and birds probably shared an ancestor among pre-dinosaur reptiles.

As Dr. Feduccia said, "This has probably been the most contentious issue in vertebrate paleontology for the past 30 years—whether birds are derived from earthbound dinosaurs or whether they are derived from antecedents of the dinosaurs."

Dr. Mark Norell, a paleontologist at the American Museum of Natural History in New York City, has made discoveries in the Gobi Desert of Mongolia that appeared to confirm the dinosaur-bird link. He said the early fossil record for birds was too spotty to support the conclusions of Dr. Feduccia's group. It is possible, paleontologists argue, that some bird not yet found in the record at about the time of archaeopteryx underwent a rapid evolutionary spurt that gave rise to the two bird lineages.

Besides, some scientists question the dates for *Liaoningornis*. At a meeting of the Society of Vertebrate Paleontology last month, Dr. Derek York, a University of Toronto geophysicist, reported new tests showing dates of 121 million years for the fossil beds in Liaoning Province.

The journal *Science* quoted Dr. Luis Chiappe, another paleontologist at the American Museum, as saying that if the new fossils are indeed substantially younger than archaeopteryx, then they may have little bearing on the origins of birds.

—JOHN NOBLE WILFORD, November 1996

Fossil Find Hints at Origin of Flight

ALTHOUGH MOST PALEONTOLOGISTS agree that birds are descendants of dinosaurs, they have yet to figure out how and why the early birds managed to get off the ground and fly. How, in particular, did the wings of birds evolve from the forelimbs of dinosaurs?

Paleontologists digging in the Patagonian region of Argentina have now uncovered fossils of what they say is the most bird-like dinosaur ever found. It was a long-legged, meat-eating creature, nearly four feet tall at the hip and more than seven feet long. It could not fly, but it had a pelvis and hind limbs like ancient birds and shoulder bones that were oriented in an avian way.

With the shoulder socket facing down and back, the animal could tuck its upper arm bones close to its body much like the wing-folding mechanism of birds. Outstretched, the forelimbs were probably capable of the full upstrokes necessary for flapping flight. It is not known if the animal had feathers.

A team of Argentinean paleontologists reported that the newly discovered dinosaur appears to be a close relative of birds and bears many anatomical characteristics of the intermediate species between dinosaur and *Archaeopteryx,* the earliest known bird. The new fossil animal has been named *Unenlagia comahuensis,* which means "half bird from northwest Patagonia" in the language of the Mapuche Indians of the region.

Dr. Fernando E. Novas, a paleontologist with the Argentine Museum of Natural Sciences in Buenos Aires, said in a telephone interview that the new fossil species "is significant because it represents a new dinosaur more closely related to birds than any other known dinosaur."

Other paleontologists said the findings added further support to theories of a close kinship between dinosaurs and birds, and should provide clues to the evolutionary transition between the two. But the critical point about *Unenlagia's* shoulder orientation was likely to provoke controversy, they said.

The discovery was announced in Washington by Dr. Novas at the National Geographic Society, which supported the research. Dr. Novas and Pablo F. Puerta, a technician at the Patagonian Paleontological Museum in Trelaw, Argentina, are reporting more details of the bird-like dinosaur in an article in the journal *Nature*.

In an accompanying article written for the journal, Dr. Lawrence M. Witmer, a paleontologist at Ohio University in Athens, said, "*Unenlagia* is a true mosaic, begging the question of where to draw the line between what is, or is not, a bird."

Dr. Mark A. Norell, a dinosaur specialist at the American Museum of Natural History in New York City, said the discovery was an important step in filling a wide gap in evolutionary history leading up to birds. "This is an active area of research," he said, and more discoveries concerning the dinosaur-bird relationship are expected soon.

More than 20 bone fragments of the bird-like dinosaur were found in January 1996 in the sandy soil of a remote Patagonian hill. Geological dating of sediments indicated that the fossil animal is about 90 million years old. That means it lived well after birds evolved; *Archaeopteryx* lived about 145 million years ago.

Unenlagia, paleontologists said, could thus be considered a cousin of *Archaeopteryx* and other extinct reptilian bids, the two lineages stemming from a common ancestor. It is a relationship comparable with that of humans and chimpanzees. Though it lived later, *Unenlagia* was probably a descendant of a lineage that diverged just before the fully feathered and flying line that leads to *Archaeopteryx* and true birds.

If its body was too big and its wing-like forelimbs too short for flight, Dr. Novas said, the newly discovered *Unenlagia* might have used its wing-flapping ability in running and jumping in pursuit of prey. A pigeon flaps its wings up and down in a similar fashion to take off from the ground.

The new evidence, Dr. Novas and Mr. Puerta wrote, could help settle a long-standing debate over the origin of powered flight by birds. Did flight evolve from the trees down, with birds jumping from limbs and gliding first, or from the ground up? The anatomy of *Unenlagia*, the scientists said, suggested that flying began from the ground up, without an intervening arboreal gliding phase.

Dr. Witmer said this interpretation of flight was certain to cause an argument among theorists.

—JOHN NOBLE WILFORD, May 1997

3

GLOBAL CATASTROPHES

Time lines show the Earth's geological history divided into discrete periods. Cambrian, Ordovician, Silurian, Devonian, Carboniferous, Permian—the stately names succeed each other like presidencies or the reigns of monarchs.

But why do periods end, their distinctive set of plants and animal species disappearing from the fossil record to be replaced in time by a new set?

A guiding principle of paleontology is gradualism, that the reason for even the most dramatic changes in the fossil record should be sought in known natural processes, like volcanoes, earthquakes, climate shift and changes in sea level.

This tried-and-true principle goes back to the earliest days of the subject, when the first paleontologists were trying to break away from the encumbrance of explanations that invoked Noah's flood and other supernatural interventions. The participants in this debate were known as uniformitarians and catastrophists, from the kinds of explanation each sought for geological change.

Time and again the uniformitarians were proved right and the catastrophists wrong. So when the physicist Luis Alvarez proposed in 1980 that an asteroid or comet impact had brought the Cretaceous period to an end 65 million years ago, paleontologists dismissed the idea as yet another manifestation of catastrophism.

But this time the catastrophists, after a bitter argument, turned out to be right. An asteroid impact is now accepted as the reason for the demise of the dinosaurs, and it becomes logical to consider asteroids, among other mechanisms, as possible causes of the other mass extinctions in the fossil record. The articles that follow discuss some of the events, catastrophic and otherwise, that divide the world's fossil history into its different chapters.

New Clues to Agent of Life's Worst Extinction

A QUARTER BILLION YEARS ago, something so terrible happened on Earth that up to 96 percent of plant and animal species were wiped out. The cause of this greatest of all global catastrophes remains unknown, but evidence is mounting that the main culprit was a monstrous volcanic eruption that covered much of Siberia with molten rock, touching off an ice age and a worldwide deluge of lethal acid rain.

The devastating Permian catastrophe swept the Earth clean of animal groups from trilobites in the ocean to ferocious reptile carnivores living on land. The "great dying," as some scientists call it, radically changed the history of life on the planet. It ushered most species into oblivion but opened ecological niches for a host of others, including the dinosaurs of the Mesozoic era that followed.

Like the other episodes of mass extinction discovered in the fossil record, the deadly wave of Permian extinctions has excited endless speculation as to what caused it. But few hard facts have turned up. One reason is that it occurred so long ago that many possible clues have been obliterated by geological processes.

Even the causes of much more recent extinctions, including the event that ended the Cretaceous period 65 million years ago and wiped out the dinosaurs, remain uncertain and disputed because of ambiguities and gaps in the fossil and geologic records. The Permian extinction, nearly four times as old as the Cretaceous extinction, is all the harder to decipher.

Nonetheless, an important new clue has emerged to what may have been a major or even principal cause of the Permian extinction. It has come from study of a geological formation called the Siberian Traps, an expanse of volcanic rock some 870 miles in diameter that borders Lake Baikal. Floods

Patricia J. Wynne

The extent of an extinction is measured by the percentage of marine animal families that became extinct; many more genuses and individual species were reduced or wiped out. Marine creatures are used as the index because the fossil record in marine sediments is far more complete than that for land animals. Families are counted because larger categories are easier to gauge. In the Permian mass extinction, the plants and animals depicted here were among the notable victims.

of volcanic basalt often have a terraced appearance, and the geological term *trap* is derived from the Swedish word for stairs.

Although the ages of the vast Siberian lava field and the Permian mass extinction were apparently similar, convincing evidence was lacking that these ages actually coincided.

An international group of scientists has now found that the eruptions that created the flood of molten basalt in the Siberian Traps occurred in something less than 600,000 years, beginning 248 million years ago. Their study also showed that the Siberian eruptions were probably violent enough to blast immense amounts of dust and aerosol droplets into the upper atmosphere, where they could have blocked sunlight for a long time, thereby radically cooling the Earth and causing polar ice caps to grow at the expense of the seas' liquid water.

There seem to have been rapid and large fluctuations in global sea levels during this period, they reported, which would have drastically upset habitats, especially of many marine organisms. Such a situation could explain why the Permian extinctions were much more severe for marine animals than land animals.

The scientists have found, moreover, that the volcanic blasts that created the Siberian Traps would have converted immense quantities of sulfate minerals into sulfur dioxide gas and sulfuric acid likely to fall back to Earth in the form of acid rain. This, coupled with climatic changes or the loss of habitats, could have snuffed out many species.

The new evidence was reported by Dr. Gerald K. Czamanske of the United States Geological Survey at Menlo Park, California, in the journal *Science,* and in a talk he gave at the American Geophysical Union meeting in San Francisco.

Dr. Czamanske, with two Australian colleagues, Dr. Ian H. Campbell and Dr. Robert I. Hill of Australian National University in Canberra, and two Russian scientists, Dr. Valerie A. Fedorenko and Dr. Valentin Stepanov of the Central Research Institute of Geological Prospecting for Base and Precious Metals in Moscow, used an ultrasensitive and carefully calibrated dating technique to gauge the ages of ancient minerals. They did this by measuring the extent to which uranium and thorium in the minerals had been transformed by radioactive decay to various forms of lead.

Some scientists, including Dr. Dewey M. McLean of the Virginia Polytechnic Institute, and Dr. Charles B. Officer and Dr. Charles L. Drake of Dartmouth College, have long theorized that vast upwellings of volcanic basalt at various periods in the past caused major mass extinctions.

In 1988, Dr. Michael R. Rampino and Dr. Richard B. Stothers of NASA's Goddard Space Flight Center said in an article in *Science* that nine of the 10 greatest mass extinction events of all time more or less coincided with continental floods of volcanic basalt in various parts of the world, and that these extinctions might have been caused by volcanism.

The most famous mass extinction, generally ranked about fifth in overall severity, was the Cretaceous event, the catastrophe that exterminated the dinosaurs. This occurred about the time of a huge eruption in what is now India, a great lava flood that created a formation called the Deccan Traps. But at about the same time, many geologists believe, one or more asteroids or comets struck Earth.

Some scientists believe that such impacts alone caused the Cretaceous extinction. Others believe the impacts were simply the final blow to Cretaceous animals and plants already weakened by environmental stresses. Another group, including many paleontologists, believes the evidence that extraterrestrial impacts caused mass extinctions is too weak to accept as conclusive.

A few experts, including Dr. David M. Raup, a paleobiologist at the University of Chicago, have argued in books and scientific papers that all mass extinctions, not only the Cretaceous extinction, probably resulted from the impacts by asteroids or comets, and that these events may occur at periodic intervals. But Dr. Raup said in an interview last week that Dr. Czamanske's volcanic explanation of the Permian extinction was exciting, if it proved to be correct.

All modern impact theories of extinction have their roots in a 1979 discovery by Dr. Walter Alvarez of the University of California at Berkeley and his colleagues. The researchers discovered a thin layer of iridium-enriched sediment separating sediments laid down during the Cretaceous period, when dinosaurs flourished, from the later Tertiary period, when dinosaurs had entirely disappeared. Iridium, a precious metal belonging to the platinum group of elements, is more abundant in meteorites than in terrestrial rocks, and many experts regard the abnormally high content of iridium in strata laid down at the Cretaceous-Tertiary boundary as strong evidence of an extraterrestrial impact.

But no such sharp iridium anomaly has been found in sediments coinciding with any of the other mass extinctions. Moreover, iridium is often found in association with volcanic eruptions, particularly in the Siberian Traps, a region known for its rich ores of platinum-group metals, including iridium. This fact has helped to keep scientific debate bubbling.

The debate took an unexpected turn at a 1992 meeting of geophysicists in San Francisco, when Dr. Rampino and a colleague, Dr. Verne Oberbeck of NASA's Ames Research Center, proposed a theory that the Permian extinctions were caused by an asteroid.

As it approached Earth 250 million years ago, they suggested, the asteroid broke up and bombarded Earth, splitting apart the supercontinent known as Gondwanaland and puncturing Earth's crust to cause Siberia's volcanic eruptions and the great Permian extinction.

The proposal was met with skepticism by most of the scientists at the meeting, including those who believe that the Cretaceous extinction, the one that killed the dinosaurs, was caused by an impact. Dr. Czamanske, among other critics of Dr. Rampino's hypothesis, said it was unlikely that even a huge asteroid could pack enough energy to cause the vast basaltic upheaval that created the Siberian Traps. Others said the geology of the Permian extinction event lacked the iridium anomaly and other features that tended to support the impact explanation in the case of the Cretaceous extinction.

"The sediments at the end of the Permian are very different from those at the end of the Cretaceous, and the two events themselves clearly had little in common," Dr. Czamanske said. "For one thing, the Permian extinctions stretched out over a million or more years, while the Cretaceous extinctions seem to have occurred much more rapidly. I think the evidence is strong that an impact by an asteroid did occur at the end of the Cretaceous, but I see no good evidence for an impact at the end of the Permian."

Dr. Czamanske, a specialist in rock formed from molten minerals, said he hit upon the new evidence of a volcanic cause for the Permian extinctions while he and his Russian colleagues were investigating possible new sources of platinum and other valuable platinum-group metals. The Siberian Traps are at the heart of one the richest platinum-producing regions in the world, and the scientists hoped to find clues that would help in spotting platinum ores in similar geological formations elsewhere.

Dr. Czamanske believes that volcanic floods like the Siberian and Deccan Traps originate as "perturbations," perhaps related to shifts in the Earth's magnetic field, deep inside the Earth, where the mantle is in contact with the core.

These perturbations are believed to launch thin jets of superhot molten rock up through the mantle, forming "plumes." As plumes approach the crust and the Earth's surface, they mushroom outward, melting surrounding rock, gathering it up and pulling it to the surface. Huge blobs of magma, some from the plume and some from the mantle through which it passed, eventually reach the surface and send vast floods of lava over the land.

Measuring the age of minerals deposited by the Siberian Traps lava flood has proved difficult, but Dr. Czamanske's colleagues isolated tiny zircon gemstones in solidified plume material. The zircons could be dated much more reliably than the lava that brought them to the surface.

By dating zircons and other minerals, and by measuring such geological benchmarks as reversals in the Earth's magnetic polarity, the scientists calculated a date for the Siberian Traps eruptions, 251 million years ago, that they regard as the most reliable ever determined. It almost perfectly matches the estimated age of the Permian extinction, a match that Dr. Czamanske and many other scientists regard as too close to be a coincidence.

A similar explanation of the Permian extinctions was proposed in 1991 by Dr. Paul R. Renne of the Institute of Human Origins in Berkeley, California, and Asish R. Basu of the University of Rochester. But their dating of the Siberian Traps was questioned as having been based on an erroneous age estimate of the mineral used as a reference standard.

"In other words," Dr. Czamanske said, "Renne and Basu came up with the right answer, but it was based on an erroneous standard of measurement, and they reached their conclusion by accident."

However persuasive the various extinction theories may seem, they are all notoriously difficult to prove. In his book *Extinction: Bad Genes or Bad Luck?*, Dr. Raup wrote: "The disturbing reality is that for none of the thousands of well-documented extinctions in the geologic past do we have a solid explanation of why the extinction occurred."

—MALCOLM W. BROWNE, December 1992

THE FIVE GREATEST MASS EXTINCTIONS

FINAL ORDIVICIAN
435 million years ago

24% of marine families lost
45–50% of marine genuses lost

Sharp changes in sea levels affected shallow tropical waters. Many trilobites, cephalopods, crinoids and other marine invertebrates wiped out.

LATE DEVONIAN
357 million years ago

22% of marine families lost
47–57% of marine genuses lost

Sea level and climate changes, possibly over 10 million years. Loss rates especially high for corals, brachiopods, placoderm fish and trilobites.

FINAL PERMIAN
250 million years ago

More than 50% of marine families lost
76–80% of marine genuses lost

Coincided with great volcanic eruptions; up to 80 percent of marine genuses and 96 percent of all species died. End of trilobites and many land species.

LATE TRIASSIC
198 million years ago

24% of marine families lost
43–58% of marine genuses lost

Depleted cephalopods; wiped out many reptiles, gastropods, bivalves and brachiopods; ended conodonts (possible fish ancestors of vertebrates).

FINAL CRETACEOUS
65 million years ago

16% of marine families lost
38–46% of marine genuses lost

Coincided with Deccan Traps and possibly with impact of asteroids or comets. Demise of dinosaurs, ammonites (shellfish) and many other animals.

Mass Extinction of Permian Era Linked to a Gas

A NEW STUDY by biologists and geochemists suggests that a huge upwelling of carbonated water from the depths of the oceans may have played a lethal role in the greatest of all mass extinctions 250 million years ago.

That mass extinction, at the end of the Permian period, reshuffled Earth's genetic deck, making way for the rise of dinosaurs in the subsequent Mesozoic era and widening the niches available for the mammal-like reptiles destined to become ancestors of the human race.

Because the events that brought the Permian period to a close occurred so long ago, much of the fossil record of the period has been erased. Scientists confront difficulties in interpreting the causes of the Permian mass extinction, which ushered in the Triassic period, the first period of the Mesozoic era, the age of dinosaurs.

One theory is based on the fact that at about the same time as the Permian mass extinction, a gigantic volcanic eruption covered much of present-day Siberia with lava. Many scientists suspect that the poisonous gases released by that event, called the Siberian Traps eruption, coupled with the climatic effects of the large-scale release of volcanic carbon dioxide, could have had deadly global consequences.

But another idea was proposed in a report published in the journal *Science* by four scientists: Dr. Andrew H. Knoll of Harvard University's Botanical Museum; Dr. Richard K. Bambach, a geologist at Virginia Polytechnic Institute and State University in Blacksburg; Dr. Donald E. Canfield of the Max Planck Institute for Marine Microbiology in Bremen, Germany, and Dr. John P. Grotzinger, a planetary scientist at the Massachusetts Institute of Technology. In their report, they examine the possibility that soda water may have been the main culprit in the Permian catastrophe.

A late Permian lowland Glossopteris forest, inundated by carbon dioxide–laden seawater, with dying aquatic creatures: trilobites (segmented animals); cephalopod ammonites (with tentacles); early ray-fin fishes (one is on its side), and an amphibian (horned creature).

On the shore, a large young herbivorous therapsid (moschops) has died; behind it, on the log, is an early cynodont, a mammal-like reptile. Its close relatives survived the Permian extinction, eventually giving rise to the mammals and humans.

Michael Rothman

81

The researchers argue that the patterns of extinctions and of the types of calcium-carbonate sediments laid down at the end of the Permian period are consistent with a global turnover of deep ocean water, which would have brought immense amounts of carbon dioxide to the surface, both dissolved in surface seawater and released as a gas into the atmosphere.

A similar association of ice ages with unusual carbonate sediments occurred at least four times during the Proterozoic era more than 570 million years ago, the scientists believe, suggesting that their theory applies to other pivotal episodes in the Earth's early history. Although the sedimentary record of the Permian extinction event has been swallowed up in many parts of the world by geological processes, well-preserved sediments from that time have been found in Japan, they said.

Although carbon dioxide is a normal respiration product of animal metabolism, too much of this ubiquitous compound is dangerous, particularly for marine animals. Carbon dioxide dissolved in blood forms carbonic acid, which acidifies the blood. Too much of it causes acidosis, which can kill animals, especially marine organisms with low metabolic rates.

Dr. Knoll and his associates do not suggest that carbon dioxide released into the air could have reached high-enough levels to have directly killed land animals. But air unusually rich in carbon dioxide could have produced a heightened greenhouse effect, trapping energy from sunlight and raising global temperatures, which would be dangerous to some animals and would destroy habitats needed by others.

Although the special conditions that may have made the global ocean surrounding the Permian supercontinent Pangea a potential killer probably do not exist in modern oceans, certain inland lakes with high levels of carbon dioxide dissolved in deep water pose threats to people and animals living nearby.

In 1986, Lake Nyos, a beautiful blue crater lake in northwestern Cameroon in Africa, killed more than 1,000 people as well as all the cattle and other animals along its periphery. There were so few nearby survivors of the catastrophe that eyewitness accounts remain sketchy, but some witnesses said a fountain that might have reached 500 feet in the air had suddenly shot from the lake, enveloping the area in a lethal mist.

Dr. Youxue Zhang of the University of Michigan at Ann Arbor and other scientists investigating the catastrophe concluded that the victims had all died of asphyxiation. Relatively heavy carbon dioxide gas erupting as a fountain from the lake displaced the air above it and deprived nearby animals and people of oxygen, they said.

Dr. William C. Evans of the United States Geological Survey in Menlo Park, California, concluded a report about the incident with the comment that "during its most violent phase the fountain would thus resemble a scaled-up geyser more than an uncorked champagne bottle."

The trigger for sudden dangerous upwellings of carbon dioxide in Lake Nyos and several similar lakes in Africa and Indonesia remains a subject of debate. Once carbon dioxide–rich water under high pressure is forced by something to rise, the pressure is released and the dissolved gas forms bubbles. The resulting foam then rises quickly, creating a chimney through the water that bursts forth as a fountain.

The ancient Permian ocean probably did not erupt with fountains of soda water but released carbon dioxide much more gradually, perhaps as small bubbles like those trickling from a glass of stale club soda.

Dr. Bambach of Virginia Polytechnic said he did not believe that the release of gas would have looked very dramatic. "It might have taken several centuries for the overturned carbon dioxide to escape," he said in an interview, "but the process could have increased the amount of carbon dioxide in the atmosphere by five or six times."

He and his colleagues believe that the injection of carbon dioxide into the air associated with the Permian extinctions occurred in two separate pulses, the first occurring two million or three million years before the close of the Permian period, and the second at the very end of the period. Besides carbon dioxide, such an upwelling of deep ocean water would probably have brought to the surface large amounts of the poisonous gas hydrogen sulfide, which would have created further survival problems for many species.

Dr. Bambach and his associates propose several possible explanations for a catastrophic carbon dioxide upwelling.

In a scenario favored by the team, the process begins with a warm global climate in which oceanic vegetation absorbs large amounts of carbon dioxide from the atmosphere through photosynthesis, then dies and deposits the carbonate as sediments in deep ocean water where there is little or no life-sustaining oxygen. The resulting depletion of carbon dioxide from the atmosphere reduces the atmospheric greenhouse effect, and global cooling takes place.

The tectonic movement of crustal plates, meanwhile, pushes up high mountains, and the upper slopes are subjected to freezing temperatures. Glaciers and ice caps form, many of them reaching the shores of the Permian supercontinent Pangea.

The glaciers begin chilling the top layer of nearby seawater, which becomes denser than the warmer underlying water. That creates an instability in which surface water begins to sink, displacing the deeper water that is heavily charged with carbon dioxide. At a certain critical point, the slowly rising carbon dioxide–rich water begins to form bubbles; that accelerates its rise, saturating water near the ocean surface with the gas and releasing carbon dioxide into the atmosphere.

Marine animals with calcium carbonate shells or skeletons build them from carbon dioxide and dissolved calcium compounds. But they are particularly sensitive to overdoses of carbon dioxide, called hypercapnia. That, the scientists believe, accounts for the extinction of countless ocean species at the end of the Permian period. Among the victims were the last of the trilobites, highly successful three-lobed arthropods that had been predators of the seas for 375 million years. Many terrestrial species also died off, including moschops, an ox-like vegetarian reptile.

Surviving the Permian mass extinction by the skin of their teeth were some of the cynodonts, mammal-like reptiles believed to have been ferocious predators with high metabolic rates. Their metabolism, Dr. Bambach and his associates say, would have helped to protect them against the effects of hypercapnia, which was fortunate for their descendants. Paleontologists believe that these animals may have been warm-blooded and that they eventually evolved into mammals.

Dr. David M. Raup, a former professor of paleontology at the University of Chicago who has championed theories explaining mass extinctions in terms of asteroid impacts, said he found the new theory of carbon dioxide upwelling plausible. "It's a good thing that someone has brought up this possibility," he said.

Dr. Bambach said that it was possible that the oceanic upheaval he believes responsible for the Permian mass extinction could have been touched off by the impact of an asteroid, although there is little evidence for such an impact. The trigger might also have been the rearrangement of tectonic plates, thrusting up mountains that created the glaciers he theorizes caused the chilling of the ocean surface water and the subsequent turnover of carbonated water. The Siberian volcanism occurring at the same time would have added still more carbon dioxide to the global biosphere.

"Whatever the cause," he said, "our model fits the patterns of extinction and carbonate sedimentation better than other hypotheses, and I think it has to be taken seriously."

—MALCOLM W. BROWNE, July 1996

Asteroid's Shallow Angle May Have Sent Inferno Over Northern Kill Zone

SCIENTISTS have reconstructed an almost blow-by-blow account of the catastrophe that overtook the Earth 65 million years ago at the end of the age of dinosaurs.

They believe a marauding asteroid bigger than Mount Everest slammed into Earth at a shallow angle, blasting an inferno of white-hot debris for thousands of miles across the young continent of North America and turning its mountains and valleys into killing fields.

The incinerated plants and animals were the first victims in a wave of global extinction that eventually erased hundreds of thousands of species from the face of Earth, marking the end the Cretaceous period.

That is the conclusion of two scientists at Brown University and the University of Rhode Island who have studied the riddle of why the global effects of the doomsday rock were so uneven.

"It was a corridor of incineration," Dr. Peter H. Schultz, the scientist from Brown, said in an interview. Dr. Schultz, a planetary geologist who specializes in impact studies, said "the kill zone went toward North America" and only later mushroomed outward around the planet to include the global effects, including "the long-term changing of the climate."

For more than 15 years, scientists have debated how the wallop of the speeding rock at the end the Cretaceous might have touched off wide extinctions of plants and animals. The main mechanism is thought to have been a global pall of dust that blotted out the Sun, wreaking environmental havoc.

But the problem with that theory is that the massacre depicted in the global fossil record is turning out to be far from uniform. Some lush parts of the planet were transformed into lifeless zones. Others were merely battered or in some cases virtually spared.

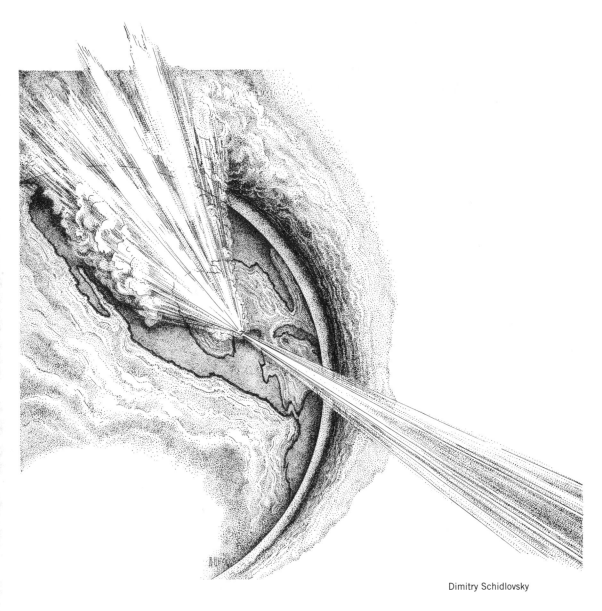

Dimitry Schidlovsky

The asteroid impact of 65 million years ago has been mapped at a level of detail that shows its major debris field blanketed North America with pulverized rock, wiping out plants and animals in its path.

The capricious nature of the catastrophe has fueled debate on whether the asteroid was really responsible for one of the biggest mass extinctions of all time, or whether other factors were at work.

Dr. Schultz and his colleague, Dr. Steven D'Hondt, a paleontologist at the University of Rhode Island, say the answer to the riddle is that the rock struck at an oblique angle, producing a colossal blowtorch of death and destruction that blazed out in one direction over thousands of miles and only later widened to create a global pall.

The key evidence, they say, lies in an impact crater found some years ago that in theory marks ground zero for the mass extinction. It is deep beneath the Gulf of Mexico and the northern tip of the Yucatán Peninsula of Mexico and is known as Chicxulub (pronounced cheek-soo-LOOB) after a nearby town. Its diameter is estimated at 185 miles.

Writing in *Geology* magazine, the two scientists say that Chicxulub's shape, as revealed by magnetic and gravity readings, is plainly characteristic of an oblique angle of approach. Moreover, they say, the global fossil and mineral record is thick with evidence of the impact's highly directional aftermath.

The mountain of rock zoomed in from the southeast, the team says, and its initial blaze of fire raced to the northwest over North America.

Reaction to the new study is positive, with scientists saying it will probably prompt new studies and new fossil hunts, if not necessarily winning over all skeptics to the doomsday theory.

"It's very interesting," said Dr. Leo J. Hickey, a paleobotanist at Yale University who has long studied the extinction riddle and was once a doomsday skeptic. "It explains a lot of disparate facts."

Dr. J. John Sepkoski, Jr., a paleontologist at the University of Chicago, said the new work "is consistent with two observations that have been problematic."

The first, he said, is that the extinction rate of fossil plants 65 million years ago was much higher in North America than in the rest of the world. The other is that grains of quartz rock that the fireball hurled around the globe are bigger in North America than elsewhere, implying that the firestorm was directional.

"It's a very interesting idea that is based on very thorough knowledge of cratering dynamics," Dr. Sepkoski said of study. But like any new idea, he added, "it has to be examined by many experts" before it becomes widely accepted.

Scientists say the study may prompt new examinations of the 150 or so impact craters that mar Earth's face, which tend to be smaller than Chicxulub. The scars might harbor signs of oblique impacts and the surrounding areas might hide fossils that show evidence of directional killing and extinction.

"People have tended to focus on global interpretations" of what happens when speeding rocks from space struck Earth, said Dr. D'Hondt.

"But to me," he said, "one of the cool things about this study is that now we have to think about regional effects" and the possibility of tongues of destructive fire.

The idea that a doomsday rock did in the dinosaurs and many other forms of life was first proposed in 1980 by a team led by Dr. Walter Alvarez, a geologist at the University of California at Berkeley. He and his colleagues had found unusually large amounts of the rare metal iridium in sediments laid down about the time the dinosaurs died out, at the end of the Cretaceous period. They proposed that the iridium came from a cosmic catastrophe when an asteroid some six miles wide struck Earth.

A weak link in the theory was the lack of a candidate crater formed at the right time and big enough to touch off mass extinctions by blotting out the Sun. The theory nonetheless won wide attention and was hotly debated, leading to scores of new studies.

Evidence soon began to accumulate that the catastrophe was temporally and geographically variable, leading some scientists to dismiss altogether the idea of a doomsday rock and a global pall of dust.

In 1981, Dr. Hickey of Yale University published a paper in *Nature,* saying that evidence from land plants was compatible with gradual, not catastrophic, change at the end of the Cretaceous. In a finding only later recognized as important, he noted that extinctions late in the age were far greater in North America than elsewhere.

More clues of directionality came to light in 1990 when Dr. Bruce F. Bohor, a geologist at the United States Geological Survey in Denver, presented an exhaustive review of grains of quartz rock that the hypothetical fireball had hurled around the globe, showing their sizes. Only later was it noticed that the grains were much bigger in North America than elsewhere, implying a skewed firestorm.

Also in 1990, scientists announced that they had at last found the candidate crater, Chicxulub, a huge irregular ring on the northern edge of the Yucatán. Geological processes like erosion and sedimentation had long ago buried all visible signs of the vast scar, which lies a mile or so beneath Earth's surface.

Like hundreds of other scientists around the world, Dr. Schultz of Brown was fascinated by this find and began studying the crater's intricacies. An expert on extraterrestrial cratering, he knew well from close study of the moon and Mars that many cosmic scars are oblong, often with long plumes of debris trailing off in one direction.

And he suspected that an intruder striking Earth at an oblique angle might well account for a doomsday event, since a shallow angle would have vaporized more of

Seabed Holds Proof of Asteroid's Devastation

Scientists penetrating through nearly two miles of sea and seabed with drilling equipment have found clear evidence of just how badly the Chicxulub asteroid shook the foundations of the planet.

Scientists aboard the *Joides Resolution,* the world's largest research ship, returned late last week from a monthlong voyage in which they lowered a long pipe to drill into the seabed 1.6 miles down. The investigation site, about 300 miles off northern Florida, was 1,000 miles or so from the Chicxulub crater, which extends beneath the northern tip of the Yucatán Peninsula of Mexico.

The *Resolution* is run by JOIDES, the Joint Oceanographic Institutions for Deep Earth Sampling, a consortium of international scientific groups.

From a depth of about 370 feet beneath the seabed, the *Resolution* pulled up a 16-inch-long muddy layer cake of distinct bands revealing the evolution of the cosmic catastrophe.

One of the expedition leaders, Dr. Richard D. Norris, a paleobiologist at the Woods Hole Oceanographic Institution in Massachusetts, said the core was extraordinary because of the clarity and comprehensiveness of its bands.

The first, and oldest, shows ancient planktonic life in the sea before the asteroid struck, the second shows a jumble of rubble from melted terrestrial rock thrown skyward by the asteroid, the third shows the rusty debris of the asteroid itself, the fourth shows a dead layer of gray clay and the last shows teeming life.

Scientists say the latter layer, two to four inches thick, will likely prove important for gaining insights into how tiny organisms and other survivors of the planetary inferno repopulated the seas. The explosion of new life occurred relatively quickly, the experts note, with the dead zone lasting about 5,000 years or so.

"You can see these really minute but beautifully preserved fossils," Dr. Norris said in an interview. "It is amazing how quickly the new species appeared."

—WILLIAM J. BROAD, February 1997

Earth's crust and rock and would have lofted more debris into the air than one that hit vertically. "It can have a profound impact on the biosphere," Dr. Schultz said, "because it transfers most of the energy into the atmosphere rather than into the Earth."

Soon, he found signs that Chicxulub was indeed elongate, based on an analysis of magnetic and gravity readings. He presented this analysis at a scientific meeting in 1994, catching the eye of Dr. D'Hondt, of the University of Rhode Island. The two decided to team up, with Dr. Schultz tying together the physical evidence, and Dr. D'Hondt the biological.

"The thing that struck me almost immediately," Dr. D'Hondt recalled, "is that the idea provided a very nice explanation of why the plant record shows such a strong response to the impact in North America."

Plant extinctions are considered some of the best indicators of global upheaval, since plants are among the most abundant and sturdiest forms of life, their seeds and spores remaining alive for long periods and traveling over long distances on wind, water and animals.

In their paper, the two scientists pull together all the clues for doomsday obliquity, including evidence from extraterrestrial craters, from laboratory experiments in the development of directional fireballs and from the Chicxulub crater itself. The gravity map of the Yucatán shows a horseshoe-shaped crater, its southeastern side deep and its northwestern side increasingly shallow.

The two scientists say the intruder slammed into Earth at an angle 20 degrees to 30 degrees above the horizontal, making the southeastern side of the crater deeper than the northwestern side. In contrast, an intruder that zoomed in from straight overhead would have an angle of 90 degrees and would leave a crater that was more circular.

As for ejecta, the paper cites not only Dr. Bohor's 1990 paper on the global distribution of quartz-rock sizes but notes that only North America has two distinct layers of ejecta, which laboratory experiments suggest is a fallout signature of an oblique impact.

Finally, the paper notes the geographic diversity of extinctions, especially among plants. They were lost at very high rates in North America and much lower rates in North Africa and Antarctica.

The trend is "strikingly illustrated," the paper says, by the araucaria, a genus of primitive cone-bearing trees with flat, scale-like needles now native to the Southern Hemisphere and grown as ornamentals in other areas. These conifers include the Norfolk Island pine from the South Pacific and the Chilean monkey puzzle. Such trees once flourished across North America but vanished from such areas right at the end of the Cretaceous.

The idea of doomsday obliquity can and should be tested, the team wrote. First, additional crater studies at Chicxulub should reveal shallow faulting to the northwest, and deep faulting to the southeast. Second, studies should continue to show asymmetrical patterns of fallout ejecta. And finally, excavations of fossils from 65 million years ago should continue to suggest that North America was hardest hit.

"What we're saying is that you can have good or bad impacts from the point of view of extinctions," Dr. Schultz said, adding, "The Chicxulub kind of angle or lower should occur in about one out of every five impacts."

And that makes for a relative low frequency of global cataclysm, Dr. Schultz said. "It means we don't have them very often."

—WILLIAM J. BROAD, November 1996

Earth bears the fossilized relics of a sample of the creatures that once inhabited it. But the fossil record, as it is called, is not some orderly list of dates and places. It is a jumble of half-erased clues and shadowy hints, from which paleontologists may labor a lifetime to extract clarity and meaning. The record is not lying around for the taking; it must be painstakingly reconstructed from mud and rock, bone and ashes.

Time is the organizing principle of any history, and paleontologists spend much effort trying to establish the dates of milestone events. Because data are so fragmentary, particularly for the earliest periods of Earth's history, a new fossil find can still change the best estimate of a date by tens or even hundreds of millions of years.

Dates in paleontology are usually revised backward in time, since a new find will show that some event occurred earlier than had been supposed. The articles in this section record a spate of recent discoveries that point to new dates for the first multicellular organisms, the first vertebrates and the first interactions between pollinating insects and flowering plants.

Fossils are usually dated from the rock strata in which they are found. The rocks are datable by measuring the extent of radioactive decay in certain isotopes they contain. Recently a new method has been developed, based on what are called molecular clocks. Present-day organisms carry related versions of many fundamental genes, which are descended from a common ancestral gene.

The rate at which genetic changes occur can be estimated, and the amount of change gives a measure of how long ago the ancestral gene existed. The clock method is still new and not well calibrated. The dates it produces are often controversial because they clash with those derived by conventional methods. But the clock methods have often proved correct, and many paleontologists have learned to take them seriously.

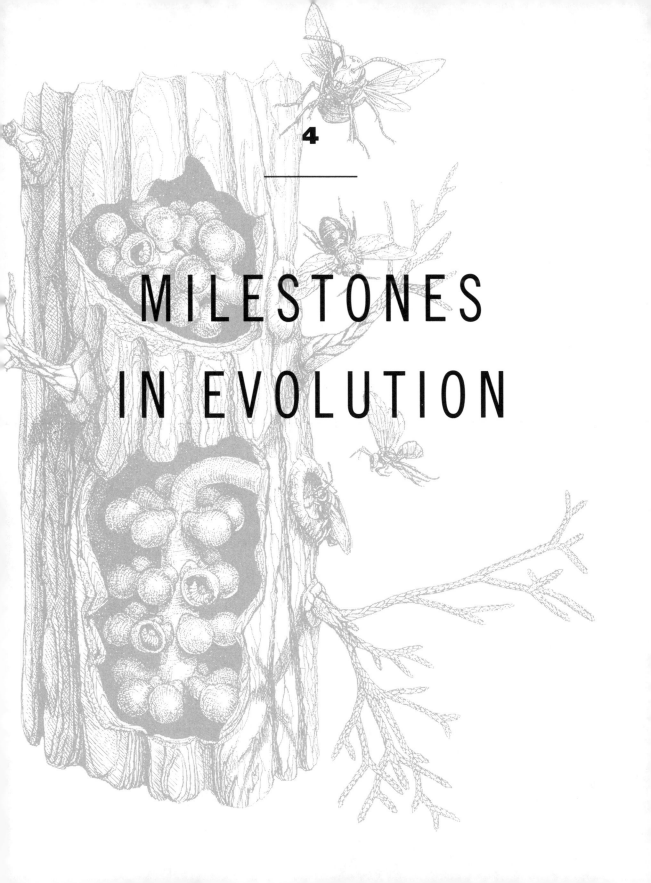

4

MILESTONES IN EVOLUTION

Three Finds Clarify Life's Murky Origins

In the billions of years of the Precambrian era, only a few simple life forms emerged, including cyanobacteria (blue-green algae), other algae and coelenterates. The Cambrian era saw many large groups of living organisms (classes or phyla) emerge, ranging from sponges and clams to snails and shrimp. New research suggests an even narrower window of time for the greatest explosion.

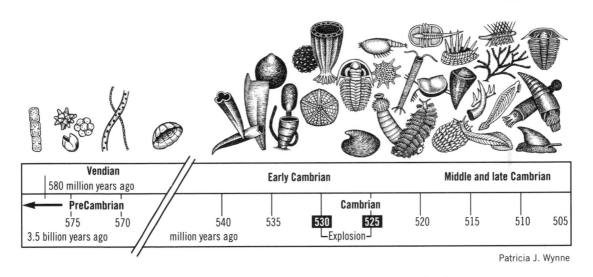

Patricia J. Wynne

LIFE, as far as scientists can tell, gained a foothold on Earth almost as soon as possible, then took an exceedingly long time rising above its simple origins and finally, 530 million years ago, erupted in a springtime of riotous proliferation. In an astonishingly brief time, insects, earthworms, corals, sponges, mollusks and animals with rudimentary backbones—all the major body plans of today—made their first appearance in what is known as the Cambrian Explosion.

But there are many yawning gaps in this early history of life, and so scientists welcomed the recent reports of three discoveries that offered important refinements in the timing of three events that have puzzled scientists trying to reconstruct the mysterious first steps in the emergence of life. The events are the introduction of large multicellular organisms, the existence of some flat jellyfish-like organisms as possible predecessors of Cambrian life and the emergence of chordates, the core group of vertebrates that would eventually include humans.

Dr. Steven M. Stanley, a paleontologist at Johns Hopkins University in Baltimore, said the findings were the latest manifestation of a "really exciting and most important activity in the field, the development of much better chronologies that constrain our views of what actually happened."

In the beginning, 4.6 billion years ago, the planet was covered with molten rock and bombarded steadily by swarms of meteorites. Not until the surface cooled down about four billion years ago could there be life, the first evidence for which are 3.8-billion-year-old fossils of a kind of blue-green algae similar to pond scum. These were simple organisms with single cells lacking nuclei. It was apparently another two billion years before more complex cells with nuclei evolved. Until recently, little evidence existed for multicellular organisms before a billion years ago.

Now, digging in sediments in northern China near Jixian, Chinese geologists say they have gathered evidence suggesting a much earlier emergence of more complex life than previously thought. They found more than 300 fossils of leaf-like multicellular plants that lived on the sea floor 1.7 billion years ago. These were described as resembling long fengshanids, which lived 700 million years later and were assumed to be the earliest reliably dated multicellular organisms.

The Chinese scientists acknowledged the discovery in Michigan of an even older spaghetti-shaped organism, but suggested that it and other early fossils "are not confidently interpreted" as multicellular plants.

Writing in the journal *Science,* Dr. Zhu Shixing and Dr. Chen Huineng of the Chinese Academy of Geological Sciences in Tianjin said the newly discovered fossils "imply that megascopic multicellular organisms originated 1.7 billion years ago or earlier." As such, they added, the new data "have implications for the understanding of the evolution and other related aspects of Precambrian life."

Another discovery may solve the mystery of what has been called a "broken link" in the poorly understood evolutionary chain prior to the Cambrian Explosion.

For half a century, scientists have not known what to make of creatures resembling jellyfish that were found in ancient sediments of the Ediacara Hills of southern Australia and subsequently in fossil beds elsewhere in the world. Were these plants or animals? Pre-cursors of later life or a failed experiment in biological innovation that came to a dead end? The problem was determining exactly when they lived and if, as it once seemed, they died out before the Cambrian Explosion and thus could not be directly ancestral to any of the new life forms.

Applying more precise dating technologies to Ediacaran fossils from the deserts of Namibia in southern Africa, geologists at the Massachusetts Institute of Technology and Harvard University determined that the youngest organisms had indeed survived into the early period of the Cambrian, 543 million years ago. The team, led by Dr. John Grotzinger of MIT, reported the results in *Science*.

"If Grotzinger and company are correct, that's excellent news," Dr. Simon Conway Morris, a paleontologist at Cambridge University in England, was quoted as saying in an accompanying article in the journal.

Scientists said the new evidence left open the possibility that there was after all no broken link in the evolutionary chain and that Ediacaran organisms could have played a role in the development of a multitude of flora and fauna that characterize the Cambrian period and are the predecessors of life on Earth today.

Dr. Samuel Bowring, an MIT geologist on the research team, said, "What this shows is that evolution likely proceeded smoothly as opposed to having a period of evolution followed by an extinction, which would open ecological niches allowing other life forms to develop."

The dating was done on grains of the mineral zircon found in trace amounts in volcanic ash. By analyzing the decay rates of uranium into lead, the geologists obtained dates for the fossil-bearing sediments that they say are accurate to within plus or minus one million years, a refinement previously unattainable on samples that old.

"Five to 10 years ago, being able to date something to within five million years was a major achievement," Dr. Bowring said. "The more precisely we can resolve time, the more sophisticated the evolutionary questions we can address."

Dr. Stanley of Johns Hopkins suggested that the Ediacaran fossil record might have been deceptive. Evidence for these soft-bodied organisms from the sea floors was found in such profusion in sediments just before the Cambrian period because nothing was scavenging on them at that time. Then the fossils seemed to disappear. Was this the sign of a true extinction, or merely an absence of fossil remains of that particular life? With the greater diversity of life in the Cambrian, he said, there could have been many scavengers munching on the Ediacara organisms before they had a chance to become fossils.

The most abundant remains of animal life originating in the Cambrian period are found in the Burgess Shale, fossil beds in the Canadian Rockies that have been the main source of knowledge about this time. But for the last decade, paleontologists have been mining an important new source at Chengjiang in the Chinese province of Yunnan, which was the site of the third discovery reported last week.

An international team of scientists described finding what may be the earliest known representative of the chordate, the branch of the animal kingdom that includes vertebrates and two lesser-known allied forms of life. Previous generations of scientists had thought that chordates evolved in the later Ordovician geological period. Later evolution, they liked to think, could imply advanced and special status to the branch of life leading to humans.

In a report in the journal *Nature,* Dr. Lars Ramskold, a paleontologist at the University of Uppsala in Sweden, and colleagues said they had identified 525-million-year-old fossils of a strange, fish-like creature, which they have named *Yunnanozoon lividum.* One of the characteristics linking the specimen to chordates is its notochord, the pre-cursor of a spinal column.

The researchers said *Yunnanozoon* appeared to belong to the division of chordates known as cephalochordates, which are closely related to back-boned animals, including humans, but not of them. Current representatives of the group include amphioxus, a shy marine creature.

The identification of the new specimen as a chordate will be controversial, scientists said. The only other possible chordate from the Cambrian period, the *Pikaia* from the Burgess Shale, has not been described in a sufficiently detailed report and so has yet to be accorded full scientific standing. But the discoverers of the Chengjiang fossil said that the presence of one division of chordates in the Cambrian period indicated that the entire branch probably existed then as well.

In a commentary accompanying the chordate report, Dr. Stephen Jay Gould, a Harvard paleontologist and evolutionary biologist, said the "unambiguously identified chordate from the still earlier Chengjiang fauna now seals the fate" of previous efforts to assert the specialness of human ancestry by separating it from the herd of new Cambrian animal forms.

"So much for chordate uniqueness marked by slightly later evolution," he wrote. "As for our place in the history of life, we are of it, not above it."

—JOHN NOBLE WILFORD, October 1995

First Branch in Life's Tree
Was Two Billion Years Ago

THE TREE OF LIFE has many branches, lichen and lily, thistle and thorn, all creatures great and small. The tree's trunk, as nearly as anyone can tell, took root under a dimmer sun in the warm waters of a cooling Earth, in the calm after its long and stormy infancy. Life began as microorganisms with a single cell, the simplest of biochemical factories wrapped in porous membrane, but at first lacking so much as a nucleus.

That occurred sometime from four billion to 3.5 billion years ago, more than half a billion years after the Earth came into being. But traces of earliest life are scant, leaving scientists at a loss to explain how life originated or when it began evolving into complex forms.

In a new effort to date some of life's critical branching events, molecular scientists have developed a "clock" showing when the major forms of life—from bacteria and protozoa to plants, fungi and animals—diverged from each other. This molecular clock is based on the rate of changes in proteins shared by different groups of organisms. The extent of these changes is an indication of when those groups separated from a common ancestor: The greater the change, the earlier the divergence.

The most significant and controversial date derived by this technique is one for the first fundamental branching in the tree of life, when eukaryotes—cells that do have nuclei packaging their genes—split off from prokaryotes—cells without nuclei. Among the estimated 10 million species today, prokaryote cells thrive as bacteria, and eukaryotes are the cells constituting nearly everything else, including humans.

Dr. Russell F. Doolittle, an evolutionary chemist at the University of California at San Diego, and his colleagues concluded from a detailed analysis of proteins of 57 different enzymes that these two basic types of living things last shared a common ancestor two billion years ago. Their findings are described in the journal *Science*.

This date is too early to suit many scientists, too recent for others. Paleontologists studying fossils had placed the divergence of prokaryotes and eukaryotes about 1.4 billion years ago. On the strength of certain genetic evidence, some of them have contended that the split occurred as early as 3.5 billion years ago, shortly after life itself began.

Early or late, the timing of the split is freighted with implications concerning the nature of life. If more complex cells appeared soon after life got a toehold, it suggests that life is a robust phenomenon and all the more likely to have evolved on other worlds, like the planets of Sun-like stars that are now being discovered. The longer it takes life to rise above the prokaryote stage, scientists suggest, the less likely it is that life is widespread in the universe.

"If Doolittle's analysis is correct, then this is rather profound," said Dr. Bruce Walsh, an evolutionary biologist at the University of Arizona in Tucson, who was quoted in an accompanying article in the journal. "It will shake things up."

Judging by initial reactions, the new date for this fateful branching of life will be fiercely debated. In particular, scientists who rely mainly on fossil studies question the reliability of such molecular clocks. Besides, they say, fossils provide no evidence for any marked change in life two billion years ago. The earliest possible fossil clues for eukaryotes are dated at 1.8 billion or 1.9 billion years ago, which would seem to be a close fit with Dr. Doolittle's results.

Dr. J. William Schopf, a paleobiologist at the University of California at Los Angeles, acknowledged that Dr. Doolittle is "very distinguished and a world-class chemist and phylogenist," but went on to say, "It seems unlikely that we can use the molecular data in the clock-like fashion Doolittle is proposing."

In the matter of timing major divergences in life, Dr. Schopf said: "There's only one court of last resort, and that is the fossil record. The fossils are there. And it is clear there is no discontinuity in the fossil record of the sort Professor Doolittle is postulating."

Even some other scientists who try to reconstruct events in early life through molecular research disputed the new findings. Dr. Jeff D. Palmer, a molecular evolutionist at Indiana University in Bloomington, questioned

Dr. Doolittle's assumption that the rate of protein change was constant, saying there was evidence to the contrary.

"It's possible that Doolittle's onto something here, but I seriously doubt it," he said.

In interviews last week, Dr. Doolittle defended his methods and results. "Many people would like to ignore this kind of molecular data, and so they criticize it on various grounds," he said. "We tried to anticipate them in designing the research and analyzing the data."

For their study, Dr. Doolittle's team examined 531 protein sequences from 57 different enzymes, encompassing 15 groups of organisms. This included nine animal groups (from fish and amphibians to birds and mammals), plants, fungi, slime mold and bacteria. Since nearly all proteins are composed of more amino acid building blocks than are nucleic acids like DNA and RNA, their "information content is so much richer," Dr. Doolittle said.

Protein molecules of enzymes were chosen, the researchers said, because enzymes have been systematically studied and are ubiquitous, occurring in virtually all living matter to operate its chemical reactions.

Although this was a more comprehensive sampling of enzymes than in any previous research for molecular clocks, Dr. Palmer objected that only one or two proteins, and none in some cases, were examined as representative of each of the major groups in Dr. Doolittle's study. "This is a very poor representation for diverse groups dating back one to two billion years," he said.

As to the criticism concerning the research team's assumption of a constant rate of change in proteins, Dr. Doolittle said the molecular clock was calibrated with known divergence events as dated by fossil studies. The number of amino acid changes that separate similar proteins in different lineages was determined, enabling the researchers to establish the rate of change as far back as 600 million years. They used this rate for fixing the time of earlier divergences.

"If you look at the protein clock too closely over too short a distance, it tends to be erratic," Dr. Doolittle conceded. "The trick is, you have to look over long periods with lots and lots of data that have been averaged. That's our contribution. We took vast numbers, by previous standards, and put them in a different way. We averaged them, as opposed to looking at them individually, and this is what has led to these results."

The most provocative dates concerned events before 600 million years ago. Until then, evolution seemed to be exceedingly slow, having produced no organisms more complex than algae, single-celled bacteria and plankton. Soon they were joined by a multitude of diverse multicellular animals, anticipating the body architecture of nearly every form of animal known today. Beginning 543 million years ago, in the early Cambrian geological period, this burst of evolutionary creativity produced the first creatures with teeth, claws, jaws and backbones. Last year scientists reported finding a 525-million-year-old fossil of the oldest ancestor yet known of the vertebrate branch of the animals.

Among the earlier branching events dated in the new research, the scientists reported, plants, animals and fungi last shared a common ancestor about a billion years ago. Fungi and animals shared a common ancestor more recently than either did with plants. All of today's bacteria stem from a common ancestor 1.5 billion years ago.

If this date for bacteria is right, it poses a new problem for scientists: understanding what kinds of bacteria might have existed as far back as 3.5 billion years ago. The first organisms are thought to have been a kind of bacteria. Paleobiologists interpret fossils as showing that the earliest known bacteria, at the base of the tree of life, resemble one form of extant bacteria, cyanobacteria.

But the new data place the divergence of cyanobacteria from other bacteria at 1.5 billion years ago. It is still possible, Dr. Doolittle said, that that common ancestor could have been a cyanobacterium that had lived since the dawn of life. In its report, his team wrote, "Questions must be asked as to what kind of organism gave rise to the present bacterial kingdom and what kinds of creatures existed before that time."

"If I knew the whole truth about these events in evolution, I would tell everybody," Dr. Doolittle said. "The fact that there are some conflicts means we need more data."

Even Dr. Doolittle's supporters, the journal noted, "agree that his protein clock may not be the last word in the book of life; but it is, they insist, a valuable first chapter."

—JOHN NOBLE WILFORD, January 1996

Using Genes as a Clock, Study Traces Animal Life Back a Billion Years

NEW MOLECULAR RESEARCH shows that the first simple forms of animal life, progenitors of creatures small and great, from beetles to humans, may have appeared on Earth more than one billion years ago, about twice as early as once widely thought.

The findings, based on chemical clues in living creatures, would mean that long before animal organisms grew bones, shells and spines that left a clear mark in the fossil record, minute creatures crawled or slithered between grains of mud or sand in primordial waters. They had no skeletons to be fossilized and uncovered by paleontologists, not even enough substance to leave detectable burrows and tracks.

But for half a billion years, these cryptic organisms presumably were slowly evolving, diverging into different lineages and finally becoming big and complex enough to leave a profusion of fossils, beginning about 565 million years ago in the late Vendian geological period and extending into the Cambrian period. The emergence then of early animal fossils was so sudden and pronounced, over a relatively short time of 20 million years, that the phenomenon is known as the "big bang of animal evolution."

A team of researchers at the State University of New York at Stony Brook made the new discovery using patterns of genetic changes as molecular clocks, a technique that is becoming increasingly fruitful in determining evolutionary history so deep in time that fossils are either absent or of uncertain value.

The scientists, led by Dr. Gregory A. Wray, a molecular biologist at Stony Brook, examined detailed records of seven different genes in different species, revealing their rates of divergence over time from common ancestors. They were thus able to estimate how long the structures of these genes had been changing. They found this to be one billion to 1.2 billion years back to the origin of animals. The earliest evidence for life, single-celled organisms resembling bacteria, is 3.6 billion years old.

The results of the analysis are reported in the journal *Science*. They support the similar findings of an earlier study in the 1980's that were viewed as intriguing but not definitive. The new work involved substantially more data and extended the putative origin of animals back another 100 million to 300 million years.

In their report, Dr. Wray and his colleagues, Dr. Jeffrey S. Levinton and Dr. Leo H. Shapiro, said the analysis of genetic divergences "cast doubt on the prevailing notion" that all the major animal groups, or phyla, "diverged explosively during the Cambrian or late Vendian, and instead suggest that there was an extended period of divergence," beginning about a billion years ago.

Other scientists described the findings as interesting and provocative, certain to stimulate debate and more research.

"It gives us a welcome new perspective on the Cambrian Explosion," said Dr. Andrew H. Knoll, a paleontologist at Harvard University. "But how much it will cause us to completely rethink the world, I don't know."

Many paleontologists, who look at the past through the lens of fossils, are expected to question the reliability of the molecular-clock methods for probing deep time.

In a commentary accompanying the report, Dr. Geerat J. Vermeij, a paleontologist at the University of California at Davis, cautioned that acceptance of the proposed ancient origin of animals hinged on the accuracy of the calibration of genetic changes in living organisms and the assumption that the rate of divergence was constant through time. Scientists are not sure such an assumption is valid.

Dr. Bruce Runneger, chairman of the earth and space sciences department at the University of California at Los Angeles, welcomed the new findings because they seemed to confirm the results of his more limited study based on blood proteins, published in 1982. "There's no doubt any more about a longer history" of multicellular animals, he said. "It's just a question of how much longer."

If the results are more or less correct, they raise important questions about what the tiny Precambrian animals were doing all that time before they developed hard body parts to leave fossils, and focus even more attention on what factors contributed to the Cambrian Explosion of fossils. What, in other words, occurred after eons of undetected evolutionary processes to trigger the spectacular rise of larger and more diversified animals at that time?

Whatever the trigger, it was responsible for a huge expansion in the size, complexity and body architecture of animals and all of the major animal phyla in existence today.

"The new work in no way diminishes the significance of the Vendian-Cambrian revolution," Dr. Vermeij wrote, "but it does separate the ecological innovations of that episode from the earlier evolution of basic animal body plans."

Since the new research indicates an exceedingly protracted evolutionary history for animal life, which never managed to rise above obscurity, the sudden change in life forms at the Cambrian revolution is all the more puzzling. But since it appears that a wide range of organisms took part in this explosion of biological creativity, Dr. Knoll suggested that the cause was probably global and environmental—specifically, a significant increase in atmospheric oxygen.

Geochemical evidence indicates that oxygen levels in the oceans and atmosphere crossed several critical thresholds as early as 900 million years ago and again close to the Cambrian period. Increased oxygen and other factors could have improved the efficiency of cells and the production of collagen, a protein that is the glue holding animal cells together. Shifting continents and drastic climate change may have been factors, too. And the half a billion years of relative stability could also have been broken by a kind of arms race.

Dr. David Jablonski, a University of Chicago paleontologist, said some scientists suspected that the introduction of predation set off the revolution. Some tiny species could have developed teeth or grasping claws.

"Once one lineage of animal life figures out how to eat another, armor becomes a matter of survival," he said, explaining that evolution by natural selection would thus favor animals with self-defense innovations leading to larger body size, protective shells and spines, and the ability to burrow and swim to flee predators.

The hard defense mechanisms that developed all of a sudden around these soft-bodied creatures allowed them at last to make their first mark in the fossil record.

—JOHN NOBLE WILFORD, October 1996

Evidence of Bone Shows Vertebrates to Be Far Older Than Once Believed

THE FIRST VERTEBRATES, the ancestral line of animals that led to the human race, appeared on Earth at least 40 million years earlier than previously believed, according to British scientists studying fossils of enigmatic, worm-like creatures called conodonts.

Their conclusion is based on new evidence that the gripping teeth of ancient conodonts were made of bone cells of the type that occur only in animals with backbones. The development of bone in primitive, soft-bodied animals was one of the great landmarks of evolution, opening the way for higher animals, starting with bony fishes and leading to the human race.

Scientists have known for years that the conodont, whose name means "cone teeth," burgeoned at least 515 million years ago in the shallow seas of the Cambrian period, in the early Paleozoic era, long before animals began living on land. But a report published in the journal *Science* offers what its authors say is the first "unequivocal evidence" that the conodont was a vertebrate, probably the first on Earth.

The conodont (pronounced KAHN-o-dahnt) is also believed to have been the world's first animal predator.

Previously, the earliest animals positively identified as vertebrates were creatures that lived in the Ordovician period 475 million years ago.

The new findings mean that major revisions will be needed in theories that try to explain how bony skeletons evolved.

Many scientists have suggested that bony skeletons had their beginnings in primitive jawless fishes with hard, mineralized plates of calcium phosphate protecting their outer skin. But the bone identified by the British group in conodonts took the form of tooth-like grippers, and there is no bone in the rest of their bodies.

Since conodonts lived in much earlier times than the first jawless fishes, one inference suggested by the British group is that cellular bone may have first evolved as a dental material, and was later adapted for use in external plates before developing in hard internal skeletons.

Paleontologists consider cellular bone a distinguishing feature of all vertebrates in any part of their bodies—not necessarily the spinal column itself. Cellular bone does not exist in any known invertebrate.

Conodonts lived from at least 515 million years ago until their mysterious extinction 200 million years ago, about the time the dinosaurs began to flourish. Although the conodont line ended, vertebrate relatives came to rule the animal kingdom, and an animal that scientists believe to have been one of the conodont's close cousins, the hagfish, survives to this day.

The hagfish, which is believed to resemble its extinct relative, is the most primitive of all living vertebrates. It somewhat resembles an eel, but has no eyes or jaw. Its circular mouth has hard, finger-like projections with which it seizes its prey.

The positive identification of conodonts as vertebrates was the work of Dr. Ivan J. Sansom and Dr. Howard A. Armstrong of the University of Durham, Dr. M. Paul Smith of the University of Birmingham and Dr. Moya M. Smith of London, an expert in the cellular structure of teeth.

"It's been clear for some time that conodonts were chordates, but this evidence that they were vertebrates is a very significant step," said Dr. Andrew Knoll, a paleontologist at Harvard University.

The fossils studied by the British team were excavated from ancient sedimentary rocks in Estonia, Greenland and Britain. The discovery of bone was made using scanning electron microscopes. These instruments were able to reveal microscopic detail in thinly sliced and polished pieces of rock cut from fossil conodont "elements," the bony grippers around the rim of the animal's mouth.

The original tissue of all fossils is completely replaced by minerals, leaving only patterns rather than the substance of the living animal. But in this case the scientists found microscopic patterns characteristic of those of living cellular bone, which is built up from successive layers of cells containing calcium phosphate and proteins.

The spine of the conodont contained no bone, but it was a stiff structure made of turgid protein cells called a notochord. Vestiges of the notochord remained in higher animals; in their early stage of development, even human embryos recapitulate the development of notochords, which are later replaced by bony vertebrae.

"You might say that this discovery of cellular bone in conodonts is the clinching evidence that they were vertebrates," said Dr. Derek E. G. Briggs of the University of Bristol, who wrote an accompanying comment in *Science*.

Dr. Briggs himself was one of the scientists credited in 1982 with the discovery of fossil remnants of conodont soft tissue. With that discovery, scientists for the first time knew what conodonts looked like.

Visible in a specimen found in sediments under the city of Edinburgh, Scotland, were the outlines of an eel-like fish. It had a finned tail and typical conodont "elements," the counterparts of teeth, in its mouth. The whole animal was less than two inches long.

Conodonts were discovered in eastern Europe in 1856, but their nature baffled paleontologists for more than a century. The fossils take the form of tiny, tooth-like cones, wedges and crowns, many smaller than one thirty-second of an inch in length.

Biologists surmised all kinds of origins for conodonts; they were variously identified as jaws of segmented worms or as parts of plants. The only agreement was that some of the conodonts were present in rock sediments corresponding in age almost to the dawn of life.

Despite uncertainties about their origin, conodonts became enormously useful to scientists, including geologists prospecting for petroleum. The ages of different types of conodonts are known quite accurately, and their presence in sediment tells scientists the geological age of that sediment.

It was also found that some conodont fossils are discolored, and the degree of discoloration depends on the amount of heat to which the fossils were exposed in the rock. Age and heating are indicators of the likelihood of finding oil in a given sediment.

—MALCOLM W. BROWNE, May 1992

Early Amphibian Fossil Hints of a Trip Ashore Earlier Than Thought

Michael Maliki

IT WAS A REAL fish-eat-fish world 365 million years ago.

That is the way Dr. Neil H. Shubin, a University of Pennsylvania paleontologist, characterizes the competitive conditions in which fishes were evolving bigger teeth, larger bodies and tougher armor, a fitness program for survival in the fierce marine world. But some clever fishes chose escape as the better part of valor, becoming the first amphibians and being able to feed in the water and flee to safety on land.

"These ancient amphibians revolutionized the whole ball game by walking away from the fray," Dr. Shubin said of this fateful transition in the history of life.

111

One of his graduate students, Edward B. Daeschler, has now discovered fossils of one of these pioneering amphibians, the oldest ever found in North America and the second oldest in the world. The discovery indicates that such transitional creatures between fishes and all subsequent backboned animals probably adapted to land earlier than once thought and swiftly established beachheads throughout the world.

The new fossils include bits of a skull and a complete shoulder of the amphibian, which has been named *Hynerpeton bassetti*. From the shoulder alone, paleontologists determined that the animal had strong forelimbs capable of supporting its body on land and moving forward and back to provide locomotion. The absence of certain muscle-attachment points on the shoulder also indicated that the amphibian had no gills. Judging by the shoulder size, the scientists estimated that animal was probably three feet long, presumably including some kind of tail.

"We have not found enough of this animal to say that it didn't have a tail that it used for propulsion through water, but we know that the forelimb was very well developed," Mr. Daeschler said. "This animal could do push-ups."

The fossils were described and interpreted by Mr. Daeschler, who is a paleontologist at the Academy of Natural Sciences in Philadelphia as well as a graduate student at the University of Pennsylvania, in the journal *Science*. His co-authors were Dr. Shubin; Dr. Keith S. Thomson, president of the academy, and William Amaral, a fossil preparator at Harvard University.

"The discovery of *H. bassetti* extends the known geographic range of early tetrapods," or four-legged vertebrates, the researchers reported, "and indicates that they had a virtually global equatorial distribution by the end of the Devonian," the geological period from 408 million to 360 million years ago.

The fossils were recovered in 365-million-year-old sandstone and mudstone sediments in north-central Pennsylvania near the village of Hyner, west of Williamsport. The way the continents were configured then, this area was just south of the equator, a marshy coastal plain of streams flowing toward an inland sea about 100 miles to the west. What is now the eastern part of North America was more or less joined to Greenland, the British Isles, Scandinavia and northern Europe.

The most complete skeletal material of an early amphibian tetrapod is the *Ichthyostega,* found in Greenland and dated 362 million years old. The earliest amphibian fossils are poorly preserved 370-million-year-old limb bones from Scotland. Other remains have been identified in Latvia, Russia, Australia and Brazil.

Since the shoulder anchors the muscles for the animal's forelimb, it is prized as a fossil for what it can reveal of an animal's mode of locomotion. In this case, the researchers were surprised by the apparently advanced development of the animal's forelimb muscles, compared with earlier amphibians and even the later specimens in Greenland, which remained better adapted to aquatic life and probably could only slither across the ground.

"Our conception of early tetrapod evolution is really changing," Dr. Shubin said, noting that the new evidence could mean that the first amphibians arose as early as 380 million years ago, not 370 million.

From an analysis of the shoulder, the researchers determined that the *Hynerpeton* had made substantial evolutionary strides from its exclusively marine past. The animal's limbs enabled it to have a sprawling stance and gait. It may even have become predominantly terrestrial by this time. The mode of locomotion suggested by the fossils differs significantly from the flexing of the backbone and sideways movement of the tail, which characterize swimming in fishes.

The details of how some fishes adapted to living on land remain sketchy. But pioneers like *Hynerpeton* seemed to live in streams and marshes, probably using their stubby limbs to maneuver in shallow water and their lungs to gulp air in the absence of sufficient oxygen in the water. They might have been drawn ashore by the tempting fare of plants and insects, which had previously colonized the land, or they might have been escaping marine predators, or both.

The new fossils raised other questions about the timing and pace of amphibian evolution. Because of the many differences between *Hynerpeton* and other early amphibians, Mr. Daeschler said, "we can see that at this point in time, there was a diversity of the types of animals experimenting in ways to adapt to living on land."

Dr. Shubin said that it was not yet clear if the advanced features of *Hynerpeton* meant that the earliest amphibians had evolved quickly to exploit opportunities open to them on land, or if the transition had begun much earlier than 380 million to 370 million years ago and proceeded slowly, though the fossils for these early stages are still missing.

In any event, it was another 25 million years after *Hynerpeton* before some amphibians evolved into reptiles, the first truly terrestrial vertebrates and the remote ancestors of birds and mammals, including humans.

—JOHN NOBLE WILFORD, August 1994

Animals Left the Water Earlier Than Thought

THE FIRST MARINE ANIMALS to colonize the land came ashore some 50 million years earlier than scientists previously estimated, recent discoveries in the English Lake District have shown.

Scientists from the British Geological Survey in Edinburgh and the University of Bristol report that the creatures that introduced multicellular life to the land environment were apparently arthropods resembling modern centipedes. While conducting a survey of rocks at several sites in northwestern England, geologists and paleontologists came across tracks left by at least one species of arthropod in the late Ordovician period 440 million years ago. The earliest previously known fossil records of land-dwelling creatures date from the Silurian period some 50 million years later.

Dr. Derek E. G. Briggs, a paleontologist at the University of Bristol, said in an interview that the newly discovered tracks pre-dated all previously known terrestrial tracks left by multicellular animals.

Dr. Eric W. Johnson, leader of the British Geological Survey team, said the rock in which the fossils were found was a greenish gray volcanic material that would have been a kind of ashy mud at the time the ancient arthropods crawled over it. Arthropod tracks were found at two sites: one in the River Licke Valley and the other in Borrowdale.

The geological formation in which the tracks were found contains evidence that the arthropods were living in fresh water that sometimes dried out, forcing them to survive on dry land. Freshwater ponds may therefore have been intermediate habitats for arthropods as they moved from the sea to land.

The paleontologists who examined the fossils were at first puzzled by the fact that some tracks showed clear impressions of individual footprints

115

while others consisted merely of blurred parallel grooves. Dr. Briggs's group shed light on the problem by experimenting with live wood lice. A louse produces distinct footprints on relatively dry mud, they found, but on wetter mud, the dragging feet leave parallel grooves like those found in the ancient mud.

"Smaller arthropods may have come ashore at even earlier times," Dr. Briggs said, "but the smaller the animal, the more difficult it is to find its traces."

Only fossil footprints and not fossils of the animals themselves have so far been found in Ordovician rocks, Dr. Briggs said, but fragments of somewhat younger arthropods have been found in rocks along the Welsh-English border and in New York State. By dissolving sedimentary rock of Silurian age (435 million to 395 million years ago) from both regions in acid, he said, scientists have been able to extract fossilized cuticles from the external skeletons of arthropods "that are remarkably similar to those of modern arthropods."

Arthropods ruled the land for some time until insects, initially flightless, gained a foothold in the early Devonian period about 390 million years ago. What then happened to the insects remains something of an enigma, although it is known that they must have thrived; the geological record contains a huge gap extending from 350 million to 290 million years ago, during which insects were not preserved as fossils.

Meanwhile, however, about 390 million years ago, another great evolutionary landmark had been achieved by the first four-limbed vertebrates to crawl on land, probably relatives of the lungfish or possibly those of the coelacanth. Both are fishes with very ancient lineages—family trees with branches that include the mammals, human beings among them.

—MALCOLM W. BROWNE, February 1995

Long Before Flowering Plants, Insects Evolved Ways to Use Them

ON A SUMMER'S DAY, nothing is more natural than the sight of bees and butterflies sticking their heads deep into garden blossoms. Insects and flowering plants are so evidently made for each other that scientists have long assumed it was the emergence of plants with flowers, or angiosperms, 125 million years ago that led to the sudden flourishing of insects in myriad forms.

This assumption, elevated to conventional wisdom and taught in biology courses, has been challenged in a new, comprehensive study of insect fossils. The greatest expansion and diversification of insects, it has now been discovered, actually began 120 million years before the advent of angiosperms. If anything, when flowering plants proliferated, insect diversification slackened.

Another surprise to emerge from the study is that today's tremendous diversity of insects does not result from the rapid creation of new six-legged species. Nor, it is supposed, does it have to do with the tongue-in-cheek observation once made by J. B. S. Haldane, a British geneticist. When asked what traits of God were evidenced by life on Earth, he thought of the 300,000-odd beetle species and remarked, "He must have had an inordinate fondness for beetles."

Instead, the study showed, the insect abundance and diversity are the cumulative effects of an extraordinarily low extinction rate. Bugs endure.

These startling findings were arrived at by two paleontologists who conducted a systematic survey of 1,263 insect families based on voluminous fossil records, including data from Russian and Chinese literature that had been overlooked in Western science. With fossils from Kazakhstan, Siberia and China thus filling in many gaps, the record for insects, starting 390 million years ago, turned out to be much richer than had been thought.

Countering the widely held view that when flowers emerged, insects radiated in profusion, a new analysis of fossils by Dr. Conrad C. Labandeira finds that flowers evolved into a world already full of insect crushers, piercers, chewers, sippers and borers. Some examples, with a contemporary insect representative, are shown here.

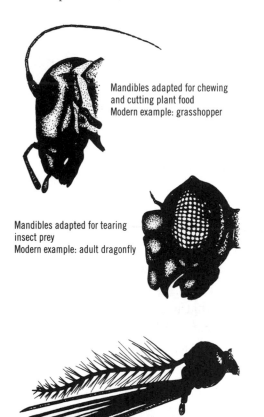

Mandibles adapted for chewing and cutting plant food
Modern example: grasshopper

Mandibles adapted for tearing insect prey
Modern example: adult dragonfly

Six stylets that pierce skin for blood
Modern example: adult mosquito

Mouthbrush used as a sieve to capture particles in water
Modern example: blackfly larva

Dr. Conrad C. Labandeira, a specialist in fossil insects at the Smithsonian Institution's National Museum of Natural History in Washington, and Dr. J. John Sepkoski, Jr., a paleontologist at the University of Chicago, described and analyzed the results of their study in an article in the journal *Science* and in interviews.

"The results contradict several notions about what macroevolutionary patterns can be seen among fossil insects and about how modern insect diversity can be interpreted," the scientists concluded in their report.

Dr. Edward O. Wilson, a Harvard University biologist and author of *The Diversity of Life* (Harvard University Press), called the work "an excellent piece of research." Although some of the findings on early insect evolution were already known, he said, "the surprise is that insects at the family level were off and running well before the flowering plants made their debut."

Dr. Leo J. Hickey, a paleobotanist at Yale University, said, "The results call into serious question some of our conceptions and preconceptions. All of us were quite comfortable with the idea that flowering plants must have had a major effect on insect diversity."

Because the results were so unexpected, Dr. Hickey said, the study and its implications are certain to be hotly debated by scientists. Dr. Sepkoski and other paleontologists at Chicago have a reputation for identifying previously unappreciated patterns in fossil histories and then playing devil's advocate in the face of orthodox thinking. Dr. Hickey said he was skeptical at first, but now considered the interpretations well supported by the fossil record in general and, particularly, by the detailed examination of the mouths of fossil insects.

The anatomical mechanisms by which insects grasp and hook, chew and grind, suck and siphon and otherwise use plants as food sources are telling clues to evolutionary changes over time. In their research, Dr. Labandeira and Dr. Sepkoski grouped insect mouthparts into 34 classes, nearly all of which are important tools in feeding off flowering plants. They established more definitively than before that with few exceptions, these were not adaptations to help insects exploit angio-sperms as a new food source, for 85 percent of these mouthpart types had evolved before the spread of flowering plants.

The basic eating machinery for insects was in place nearly 100 million years before angiosperms appeared on the scene, the scientists reported. The implications, Dr. Sepkoski said, are that angiosperms may have evolved to take advantage of insect types already in existence, not vice versa, as had been supposed.

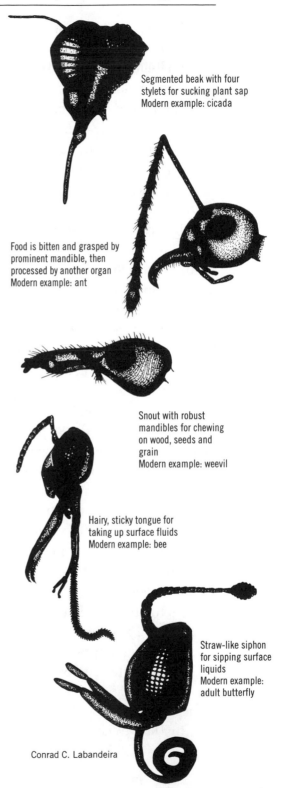

Segmented beak with four stylets for sucking plant sap
Modern example: cicada

Food is bitten and grasped by prominent mandible, then processed by another organ
Modern example: ant

Snout with robust mandibles for chewing on wood, seeds and grain
Modern example: weevil

Hairy, sticky tongue for taking up surface fluids
Modern example: bee

Straw-like siphon for sipping surface liquids
Modern example: adult butterfly

Conrad C. Labandeira

Insects were the first animals to adapt to life on the land and in the air and became such an evolutionary success that they are the most diverse class of animal life on Earth today. Several hundred thousand species have been identified, including ants and bees, beetles and crickets, dragonflies and earwigs, fireflies and grasshoppers and such pests as fleas, lice, termites and mosquitoes. Nearly all insects have six legs, one pair of antennas, one or two pairs of wings and three distinct body parts—the head; the thorax, where the legs and wings are attached, and the abdomen.

As delicate and fragile as insect bodies are, it is remarkable that enough of them have been preserved to trace insect evolution. Some of the most revealing remains are encapsulated whole in amber, the hardened tree resin where the bugs had been trapped. Of the multitudes of insects that were engulfed in sediments, enough left their immortal traces, their soft bodies flattened like the impressions of leaves on ancient stone, so that scientists could reconstruct their histories.

In one important case, the Lepidoptera family of moths and butterflies, Dr. Labandeira and Dr. Sepkoski said their fossil remains prior to angiosperms were too sparse to draw conclusions on their long-term history of diversification.

The earliest known insect fossil is a relative of the silverfish called a bristletail that lived 390 million years ago, 150 million years before dinosaurs began their spectacular rise. Dr. Labandeira made the discovery in a piece of mudstone in 1988 while a graduate student at Chicago. Because of the tiny specimen's similarity to some modern bugs, scientists said, the discovery suggested that insects had already undergone considerable evolution over millions of years before this creature lived. Consequently, the transition of animal life from the sea to the land must have occurred earlier than previously estimated, perhaps before the end of the Silurian geologic period, which closed 408 million years ago.

As the new study established in greater detail, insect evolution took off about 325 million years ago with a revolutionary innovation: wings. With their flight apparatus, many insects were able to disperse to the far ecological corners, across deserts and bodies of water, to reach additional food sources and inhabit a wider variety of promising environmental niches.

In *The Diversity of Life,* Dr. Wilson noted that Sir Richard Southwood, a British biologist, attributed the pre-eminence and diversity of insects to their size and the phenomenon of metamorphosis, as well as the invention of wings. Being very small, for example, insects have been able to fill a variety of ecological niches that could not sustain larger forms of animal life; this leads to greater diversity, which in turn is a defense against extinction. Likewise, the metamorphosis from one life stage to another, from larva or nymph to adult, allows the species to penetrate several habitats, improving their chances for success.

To these sources of insect success Dr. Wilson said he would add the advantages of pre-emption. "Because insects were the first to expand into all the terrestrial niches, including the air," he wrote, "they were no doubt too well entrenched to be evicted by newcomers."

The only severe disruption in insect evolution came as a result of the greatest mass extinction of all time, the one between the Permian and Triassic periods 250 million years ago. That extinction wiped out 65 percent of insect groups. If this insect extinction was the result of widespread deforestation, as some scientists suspect, Dr. Sepkoski cautioned that the current destruction of forests, particularly in the Tropics, may be creating conditions similar to those that precipitated the Permian devastation.

Some readings of the fossil record show that mass extinctions seem to occur about every 26 million years. Insects have flitted through the others, even the one at the end of the Cretaceous period, which killed off the dinosaurs 65 million years ago, with only minor losses. According to the new study, 84 percent of the insect families living in the Cretaceous period are still around. By contrast, less than 20 percent of the vertebrate families living then have survived to the present.

Since the sharpest rise in insect diversity began soon after the Permian extinction, this presumably is a result in part of co-evolution, or the measurable response of one type of organism to another, in this case an interaction between insects and plants. Not flowering plants, however, but the non-flowering kind, such as ferns, cycads and conifers, that colored the world green before angiosperms came along 125 million years ago.

Dr. Labandeira and Dr. Sepkoski concluded that it was these seed plants, known as gymnosperms, and not angiosperms, that "provided the stage for the spectacular evolutionary history of insects." Moreover, they asserted, "the more startling interpretation that can be drawn from the data is that the appearance and expansion of angiosperms had no influence on insect familial diversification."

Paleobotanists are not sure, but the first flowering plants may have resembled magnolias or possibly tiny herbs like the black pepper plant or else an aquatic plant known as the hornwort. In any case, angiosperms are the dominant plants today. But most of the insect mouthparts for eating, though now used to exploit flowering plants, evolved to handle pre-angiosperm consumption, chewing on leaves and seed, boring into stems or siphoning aquatic invertebrates.

The scientists said they concentrated their study on family groups, rather than species, because it takes a sweeping evolutionary change to produce a new family. A new species could arise as a response to modest or isolated environmental challenges that do not necessarily reflect significant changes in living conditions.

Dr. Peter Crane, a paleobotanist at the Field Museum of Natural History in Chicago, said the new research "fits nicely with my own ideas," which are based on observations that the angiosperms were not the first plants to benefit from insect pollination. His studies, he said, show that insects had developed close relationships, which included pollination, with plants that came before angiosperms. One notable example is the Bennettitales, a close relative but not direct ancestor of angiosperms.

As surprising as the new findings are, Dr. Wilson said he would not characterize them as revolutionary. Flowering plants may not have produced a sharp increase in insect diversity, which had been biological dogma. But he emphasized that in many other ways it was still true that flowers and bees were meant for each other, as anyone can see on a summer's day.

—JOHN NOBLE WILFORD, August 1993

In the Triassic period, plants carpeted the ancient continents. Gymnosperms, ancestors of today's plants, evolved naked seeds whose pollination relied mostly on the wind. The advent of flowering plants, angiosperms, with their more efficient pollination, allowed the newcomers to colonize and dominate the Earth.

The Triassic Period
About 200 million years ago

Paleobotanists suggest that the earliest flower could either have been complex, with all sex parts in one blossom, like the magnolia, or simpler, with male and female organs divided among different flowers; this type of early flower may be an ancestor of the black pepper or water lily.

Magnolia

Water lily

Petals

Ovaries

Modern flower

Michael Rothman

123

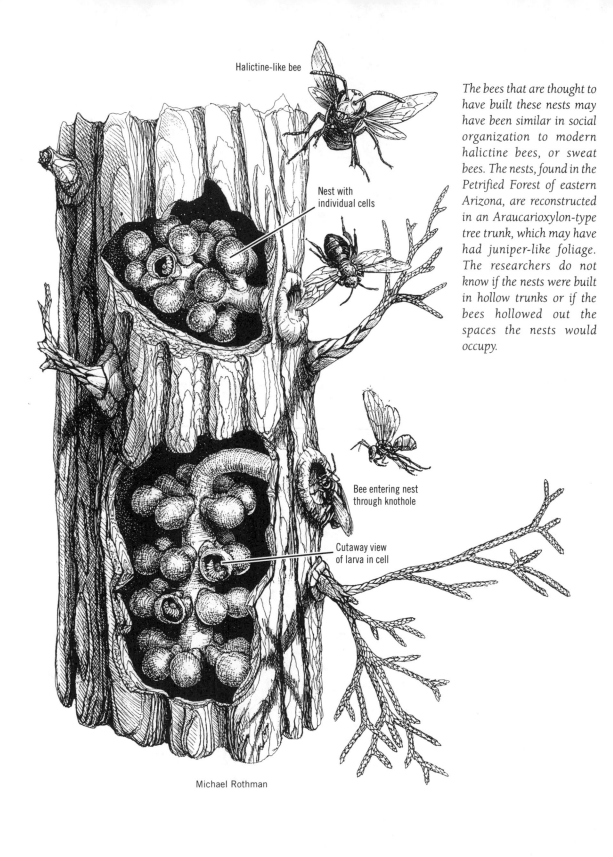

Halictine-like bee

Nest with
individual cells

The bees that are thought to
have built these nests may
have been similar in social
organization to modern
halictine bees, or sweat
bees. The nests, found in the
Petrified Forest of eastern
Arizona, are reconstructed
in an Araucarioxylon-type
tree trunk, which may have
had juniper-like foliage.
The researchers do not
know if the nests were built
in hollow trunks or if the
bees hollowed out the
spaces the nests would
occupy.

Bee entering nest
through knothole

Cutaway view
of larva in cell

Michael Rothman

Which Came First: Bees or Flowers?
Find Points to Bees

BEFORE TIME turned them to stone at least 220 million years ago, the fossilized logs of the Petrified Forest in eastern Arizona stood as tall trees in a tropical environment. Many of them, scientists have now discovered, still bear traces of insect nests the trees once harbored. The logs are riddled with holes containing little chambers strung together in lines or clusters, nearly everything about them resembling the nests of modern bees.

The problem is that flowers date from only half as long ago. Could bees have lived before flowers? The very idea, once unthinkable, is upsetting traditional theory about the early history of bees and their supposed co-evolution with flowering plants, or angiosperms.

If confirmed by further research, the new findings at the Petrified Forest mean that bees were buzzing around some 140 million years earlier than previously thought. The oldest known fossil of a bee is an 80-million-year-old specimen trapped in amber from present-day New Jersey. Scientists now must be on the lookout for fossil bees to fill that huge time gap.

And then they must figure out what those bees were doing before the emergence of angiosperms, the earliest evidence for which is dated at 120 million to 110 million years ago. Either flowers actually appeared much earlier than anyone can conceive, or the first bees did without flowers for a long time, feeding on and pollinating cone-bearing, woody plants known as gymnosperms, a group that includes conifers, cycads and ferns.

In the latter and more likely case, scientists said, the discovery casts serious doubt on the standard theory that flowering plants and social insects like bees more or less evolved together, with the spread of flowers presumably influencing the development and proliferation of the bees.

"This new evidence suggests it was probably the other way around, and that insects like bees and wasps may have facilitated the evolution and diversification of angiosperms," said Stephen T. Hasiotis, a paleobiologist at the United States Geological Survey in Denver and a doctoral student in geology at the University of Colorado at Boulder.

Mr. Hasiotis came upon the fossil bee nests while conducting studies aimed at reconstructing the ancient ecosystem and climate at the Petrified Forest, particularly as they applied to invertebrate life. He and other researchers found the remains of several hundred nests and cocoons, and tests put their ages at 220 million to 207 million years. Although no bee or wasp body parts were found with the fossil nests, they said, the only creatures that make similar structures today are bees and wasps.

The discovery was announced last week by Mr. Hasiotis at a regional meeting of the Geological Society of America, held at Montana State University in Bozeman. His collaborators were Dr. Russell Dubiel of the Geological Survey in Denver and Dr. Tim Demko of Colorado State University.

Despite the discovery's stunning implications, other scientists tended to react favorably, in part because the evidence seemed compelling and it supported recent revisionist thinking about insect evolution. This stems from a growing recognition that the greatest expansion and diversification of insects occurred many millions of years before the appearance of flowering plants.

"We're all very impressed," said Dr. Charles D. Michener, an entomologist at Kansas University in Lawrence who is the author of *The Social Behavior of the Bees,* published by Harvard University Press in 1974. Mr. Hasiotis visited the Kansas campus earlier this year and showed the evidence to Dr. Michener and his colleagues.

Dr. Michener agreed that the fossil nests looked like the clusters of chambers, or cells, that make up the nests of modern bees. But like other scientists, he cautioned that more research would be needed to confirm the findings. It is always possible that some insect no longer extant made bee-like nests back then. The best evidence, of course, would be to find some fossil bees associated with the nests, he said.

"It's exactly what we would have expected," was the reaction of Dr. J. John Sepkoski, Jr., a paleontologist at the University of Chicago. It was he and Dr.

Conrad C. Labandeira, a specialist in fossil insects at the Smithsonian Institution's National Museum of Natural History in Washington, who reported the results of a comprehensive study of insect evolution. Their conclusions challenged orthodoxy by pointing out that the appearance of flowering plants did not prompt the great diversification of insects because they had been flourishing at least 120 million years before that.

Mr. Hasiotis found the ancient nests over the last four years of fieldwork among the fossil logs. They were usually in shallow hollows inside the trunks, reached through knotholes. Each nest consisted of 15 to 30 cells shaped like little flasks, each less than an inch long. Each shell had a narrow entrance that led to a wider chamber. The walls were probably constructed out of sap and resin from the tree.

In shape and size, Mr. Hasiotis said, each cell was "virtually identical" to similar enclosures in the nests of modern bees. Their practice is to deposit an egg in each cell, along with a store of pollen and resin as food for the larva that emerges from the egg.

The researchers also found the remains of other nests underground and clusters of cocoons in the logs. They closely resemble cocoons of present-day wasps. The evidence could push back the known origin of wasps by as much as 100 million years.

"The nest construction of these insects is fairly complex, indicating a highly developed behavior," Mr. Hasiotis said.

The nest builders would not have been the same species as modern honeybees, he said, but probably primitive insects related to a smaller species today commonly known as the sweat bee. The appearance of the nest suggests the builders would have had an anatomy similar to that of today's bees and wasps. That is, they would have needed a flexible and jointed head, thorax and abdomen with strong legs and the ability to fold the wings behind and flat against the back in order to maneuver around individual nest cells.

Because it is difficult to imagine bees without flowers, the new findings could have enormous implications for botany.

As it stands now, a tiny fossilized plant that lived 120 million years ago is considered the oldest known flower. Identified by Yale University botanists in 1990, it was apparently an herb, barely one inch tall and resembling the black pepper plant. Only later, it is assumed, did angiosperms evolve more dazzling flowers and nectar to lure insects, birds and bats to act as agents, transferring the pollen of one plant to the seed of another.

But recent research has led some botanists to suspect that the first angiosperms probably originated much earlier—perhaps as far back as 200 million years ago in the Triassic period, when dinosaurs were getting started. If it could be established that bees already existed then, theories of an earlier history for angiosperms would receive a big boost.

But the earlier bees would not necessarily have required angiosperms, said Dr. Thom Taylor, a paleobotanist at Ohio State University in Columbus. Although he has not studied the evidence for fossil bee nests, he could conceive of bees existing in a green world of ferns, conifers and cycads and other ancient gymnosperms, the dominant vegetation in the Triassic.

The pollen of these plants is normally scattered by wind, not by insects and birds. But Dr. Taylor said, "It would not surprise me to learn that bees and fern plants developed a relationship that involved pollination."

In that case, as Mr. Hasiotis speculated, flowering plants did not so much open up new ecological niches for insects, including bees and wasps. Instead, they may have evolved to compete for the attention of insects that were already flourishing.

"Primitive angiosperms probably took advantage of this bee and wasp behavior by developing various colors of flowers as a pollination strategy to compete with gymnosperms," the researcher said. "Over time, most of the insects actually shifted from gymnosperms to the flowering plants."

Angiosperms may well be late bloomers as far as bees are concerned, but with the help of a lot of busy bees they have certainly flourished.

—JOHN NOBLE WILFORD, May 1995

5

STORIES FROM
THE FOSSIL
RECORD

istorians deal with the period of the world's history for which written records exist, a mere 5,000-year span. Life has existed on Earth for maybe 3,800,000,000 years, a period 760,000 times as long as written history. The paleontologist's task is to reconstruct the unwritten history of the planet.

The difficulties are obvious. Only a small and random sample of all the creatures that have lived on the Earth have been preserved. For an animal to be fossilized, an unusual combination of circumstances must occur, typically including the immediate burial of the corpse followed by years of freedom from any further disturbance while the tissues are replaced or fortified by minerals.

Given these requirements, it is amazing how much of the lost past the students of fossils have been able to retrieve. Sometimes these reconstructions are little better than educated guesses, as when an entire animal is drawn on the basis of a toe bone, but for the most part paleontology is a science that builds on accumulated knowledge.

It is, in particular, a fresh and vital science, despite the antiquity of its subject matter. New and ingenious methods of extracting information from ancient rocks are continually being invented. New locales rich in fossils are discovered. Old mysteries are solved, only to be replaced with new ones, as the following articles show.

Reconstructing a Galloping Crocodile After a Mere 200 Million Years

ABOUT 212 MILLION YEARS AGO, a small galloping crocodile roamed the floodplain that is now the Connecticut River Valley. The bones of nearly all of the creature's vertebrate contemporaries in the region have disappeared, but a three-inch skull of the little crocodile has turned up in exposed rock in a road cut.

The discovery was made near Cheshire, Connecticut, in March 1995, but it has taken more than a year for scientists to classify the skull with certainty and to dislodge the delicate bone from the hard rock in which it was embedded. The find was announced during a news conference at Dinosaur State Park by the paleontologists studying the fossil: Dr. Paul E. Olsen of the Lamont-Doherty Earth Observatory of Columbia University, Dr. Hans-Dieter Sues of the Royal Ontario Museum in Toronto and Dr. Mark A. Norell of the American Museum of Natural History.

Part of the interest in the skull stems from the rarity of vertebrate fossils of any kind from the Mesozoic era in the northeastern United States. Many tracks of dinosaurs and other animals of the Mesozoic era have been found in Connecticut (particularly around Dinosaur State Park, south of Hartford), but bones and fossils of the animals themselves have mostly disappeared, Dr. Olsen said in an interview.

The little crocodile discovered by Dr. Olsen, which has not yet been given a formal name, seems to be closely related to a similar fleet-footed Mesozoic crocodile called *Erpetosuchus* that was found in Scotland a century ago. Only the one *Erpetosuchus* has ever been found, Dr. Sues said in an interview.

While the Connecticut skull consists of actual bone, he said, the Scottish specimen, which was found in 1894 near Elgin, Scotland, is only an

131

imprint of parts of the animal in soil that eventually solidified into coarse sandstone. The reptile, about two feet long, had long, graceful legs with four toes on the hind limbs and five toes on the forelimbs, and it had teeth only in the front of its jaws. The Scottish impression shows that it also had armored skin plates.

"We used to think that those plates were for defense," Dr. Sues said, "but it now looks as if they helped to support vertebrae assisting in the animal's upright stance."

Although the Connecticut find consists only of a partial skull, its similarity to the Scottish specimen led to the conclusion that the animals were practically identical.

"This little reptile walked upright on its four legs, not with its legs splayed out, as is the case with modern crocodiles," Dr. Olsen said. "It could truly gallop, with all four feet sometimes off the ground simultaneously. We know that some present-day crocodiles in Australia can reach speeds of 15 miles an hour, so this Triassic-period crocodile could probably reach even higher speeds. It probably spent all or most of its time on land."

By the time the galloping crocodiles appeared in the mid-Triassic period, small dinosaurs had evolved, but had not yet achieved the ecological dominance they reached in the later Jurassic and Cretaceous periods. Unlike the distantly related dinosaurs, the crocodile lineage has survived to the present day.

In the Triassic period—the first period of the Mesozoic era—all the continents were joined in the supercontinent known as Pangea, so that crocodiles, like other animals, could travel over land between what is now Scotland and what was then a geological formation called the Hartford Basin, which encompassed today's Connecticut River Valley.

Dr. Olsen said that the Hartford Basin's fine-grained red sandstone— the same stone used in New York City's "brownstone" houses—had been accurately dated to an age of 212 million years. Dr. Olsen had chosen the road cut through the rock formation near Cheshire to look for fossils because of its age, he said. The road cut itself was excavated when the road was built about six years ago.

"There are lots of little white mineral flecks in the stone," he said, "and at first I wasn't sure about one particular little white fleck that looked like a fossilized root but could also have been bone. I came back another day for a second look. It was my wife, Annika Johansson, a graduate student at Lamont-Doherty, who realized that it was a skull. The white fleck we were looking at was the animal's snout."

—MALCOLM W. BROWNE, November 1996

Legged Snakes?
First Reliable Evidence Is In

A drawing of the snake that two pale-ontologists say provides compelling evidence that long ago, snakes had legs. The evidence came from a reappraisal of fossils of the snake, found in Israel.

Tom Saunders

PALEONTOLOGISTS say they have found the first compelling fossil evidence that long ago, snakes grew legs.

134

In a reappraisal of fossils from a limestone quarry in Israel, paleontologists identified specimens, previously thought to be a lizard species, as the most primitive known snake—so primitive that it still has short but well-developed hind limbs. This slender, three-foot-long snake, *Pachyrhachis problematicus,* lived in a shallow sea 95 million years ago.

The discovery, reported in the journal *Nature,* could be a significant step in determining the origin of snakes, which has been obscured by a frustratingly skimpy fossil record, and tracing their evolutionary history. It enables scientists not only to prove that early snakes did have legs, like other reptiles, but perhaps to establish more clearly their relationship to the wider order of lizards, one of the unsolved mysteries of evolution.

Questions about whether snakes ever had limbs and how they might have lost them have long intrigued scientists. Ancient people must have wondered, too, which would account for the story of the most famous snake of all, the source of temptation in the Garden of Eden.

When the serpent beguiled Eve into tasting of the tree of knowledge, according to the Book of Genesis, God condemned the serpent to go on its belly "and dust shalt thou eat all the days of thy life."

The two fossil specimens that Dr. Michael W. Caldwell of the University of Alberta in Edmonton and Dr. Michael S. Y. Lee of the University of Sydney in Australia examined last year at Hebrew University in Jerusalem had two hind legs, each only a little more than an inch long. And these were not lizards, as they had been classified 20 years ago, when they were excavated at the Ein Jabrud Quarries, 12 miles north of Jerusalem on the West Bank.

In their journal report, the two paleontologists said the small, narrow, lightly built skull of *Pachyrhachis* had many characteristics found only in snakes. The braincase is fully enclosed in bone. Other bones and the jaws are loosely connected, which gives snakes the widemouthed flexibility to swallow whole their prey of frogs and rodents. Even the number and nature of the vertebrae seemed to stamp the specimens as snakes.

Dr. Caldwell and Dr. Lee called this "compelling evidence" that *Pachyrhachis* was "a primitive snake with a well-developed pelvis and hind limbs."

In a telephone interview, Dr. Nicholas C. Fraser, a paleontologist at the Virginia Museum of Natural History in Martinsville, agreed that this "is unquestionably a primitive snake." Dr. Fraser wrote an accompanying article in *Nature* commenting on the research.

Dr. Fraser said the scientists would probably meet with considerable resistance, and require much more evidence, before they could win over paleontologists to their second conclusion—the link between snakes and a particular group of marine lizards—based on the analysis of the fossils.

Since the snake fossils were found in marine sediments and so must have lived in the sea, Dr. Caldwell and Dr. Lee looked for and found anatomical characteristics linking the creature to a group of extinct marine lizards known as mosasauroids.

Most of these lizards were smaller than *Pachyrhachis,* though the best known of this group was the *Mosasaurus,* a huge sea monster with jaws three feet long. This group evolved later. Modern members of the group include monitor lizards like the Komodo dragon.

The American paleontologist Edward Drinker Cope proposed an evolutionary link between snakes and mosasauroids more than a century ago, but other scientists would have none of it. Then, as now, the conventional view was that snakes probably evolved from a group of burrowing lizards, which lost their limbs to facilitate their underground activities. The most primitive snakes known then were burrowers, like the blind snakes.

Still, paleontologists were prepared to accept the idea of limbs in the distant past of snakes; after all, all living boas have vestiges of the hind limb. But many of them expressed surprise over the possible marine origin of snakes.

Dr. Caldwell said in an interview that the evidence "supports previous ideas by Cope that snakes originated in marine environments and is the first plausible alternative" to the hypothesis linking snakes with burrowing lizards. But he emphasized that the nature of the link with the mosasauroids was not clear.

"All modern and fossil snakes share a common ancestor with the mosasauroids," he said. "But we don't know how far back that is. As for the question of when snakes lost their limbs, that one is still up for grabs."

Though Dr. Fraser foresaw debate over the snake-mosasauroid connection, he said, "I think they will eventually be proven to be correct."

But this, he said, would require more fossil evidence comparing early snakes with their contemporary mosasauroids.

—JOHN NOBLE WILFORD, April 1997

Galapagos Mystery Solved: Fauna Evolved on Vanished Isle

TO BIOLOGISTS, the Galapagos Islands are very special places, not least because their distinctive wildlife prompted Darwin to hit upon the theory of evolution. But the islands, off the coast of Ecuador, have long posed a troublesome evolutionary mystery: Geologically, they are far too young for evolution, operating at its usual pace, to have produced the many unique life forms that now inhabit the islands.

A team of oceanographers and geologists has now happily resolved the mystery. They have found evidence that the first plants and animals to colonize the Galapagos chain probably landed on ancient islands that are no longer visible because they long ago sank beneath the waves. Their inhabitants were presumably forced to move on to the younger islands that exist today.

The discovery confirms a controversial hypothesis by two molecular biologists, Dr. Vincent M. Sarich and Dr. Jeffrey S. Wiles of the University of California at Berkeley. In 1983 they predicted that such "drowned" islands would be found. They reasoned that only the existence of long-vanished islands could account for the extensive evolutionary changes undergone by Galapagos species in the period since their ancestors arrived on the islands, which were originally lifeless. Plants and animals are believed to have arrived on the Galapagos aboard seaweed mats or driftwood rafts from the South American continent.

In the journal *Nature*, a team of researchers say they have found clear evidence that the oldest Galapagos Islands simply crumbled beneath the waves, a conclusion that strongly supports the Sarich-Wiles hypothesis. The authors of the report are Dr. David M. Christie of Oregon State University and his colleagues at the Universities of Oregon and California, Cornell University and the National Oceanic and Atmospheric Administration.

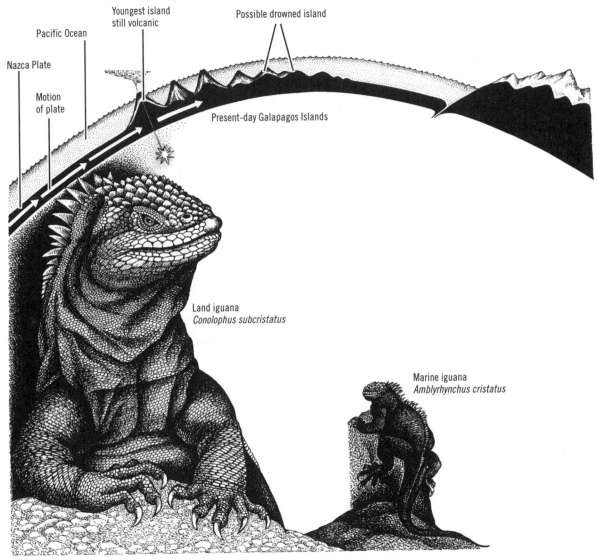

Nazca Plate

Pacific Ocean

Motion
of plate

Youngest island
still volcanic

Possible drowned island

Present-day Galapagos Islands

Land iguana
Conolophus subcristatus

Marine iguana
Amblyrhynchus cristatus

Patricia J. Wynne

New evidence from the sea floor suggests that the Galapagos chain once included older islands that have now sunk beneath the waves. On such islands the chain's unique species, like the land iguana and marine iguana above, would have had much longer to evolve from a common mainland ancestor than was previously thought. The island chain was created as the Nazca Plate moved over a "hot spot." The oldest island, now sunk, was 370 miles closer to the mainland than are the present-day islands.

In a 26-day expedition, the scientists made detailed soundings and dredged samples from undersea mountains along the Carnegie Ridge, a submerged crest extending along the Pacific Ocean floor east of the Galapagos toward the coast of South America.

"All along this ridge, and especially around the seamounts protruding from it, we dredged up round basalt pebbles and cobbles," Dr. Christie said in an interview. "We know of no geological process other than beach erosion and near-surface wave action that could produce these rounded forms. The debris that settles to the ocean floor from underwater volcanoes is very different from these. We therefore conclude that these seamounts at one time reached above the surface as volcanic islands, and were later eroded away."

The submerged seamount of the Galapagos chain lying closest to the present-day coastline of South America is about 370 miles west of Ecuador, something less than half the distance from Ecuador to the existing Galapagos Islands. The age of this seamount, whose summit is now some 6,500 feet below the surface of the waves, is about nine million years, the scientists determined. That age is much greater than the estimated two million to three million years of today's Galapagos Islands.

The existence of an island chain for nine million years would be long enough to account for the state of evolution of the Galapagos animals seen today, biologists say. Moreover, according to Dr. Christie, there are probably still more ancient members of the Galapagos archipelago that remain to be identified, and they may be 90 million years old.

Dr. Sarich and other molecular biologists have demonstrated a chemical basis for the ticking of an evolutionary clock at a more or less constant rate. Essentially, each tick occurs when one amino acid in the backbone chain of a particular protein molecule is switched for another. The protein Dr. Sarich uses for his clock is albumin, and he reckons that in a typical species, between 2.5 and 3 of these substitutions occur in the course of a million years.

He and his colleague, Dr. Allan C. Wilson, now deceased, estimated that because the proteins of chimpanzees and human beings differ by 12 units, the two species must have diverged from their common ancestor about five million years ago.

Dr. Sarich also extensively studied the protein chemistry of several Galapagos species, particularly that of the marine and land iguanas. These two species, Dr. Sarich said in an interview, clearly descended from a common ancestor, a seafaring pioneer that floated from the South American coast aboard some kind of natural raft all the way to the Galapagos.

Reptiles are able to withstand long periods without water, he said, and this may have made it easier for iguanas to survive immense ocean journeys from South America to the Galapagos, and even to Fiji and Madagascar. But many other species—plants, birds and mammals, as well as the ancestors of the famous Galapagos giant tortoises—must have occasionally survived the ordeal of raft voyages lasting weeks or months from the mainland, carried westward by the Humboldt Current. Having arrived at the islands, they began evolutionary trees of their own that were different from those of their mainland cousins.

The marine and land iguanas of the Galapagos are more closely related to each other than either is to mainland relatives. However, Dr. Sarich said, they have evolved in very different ways. The units of difference in the amino acids of their respective albumins suggest that these two species must have diverged from their common ancestor many millions of years ago, and this finding led Dr. Sarich and a co-author to publish a paper in 1983 provocatively entitled: "Are the Galapagos Iguanas Older Than the Galapagos?"

"So you can see why," he said, "we were pretty sure these sunken islands would eventually turn up. I'm not at all surprised by Dr. Christie's discovery."

This view was by no means universally accepted. In 1985, Dr. Carole S. Hickman, a paleontologist at the University of California at Berkeley, and Dr. Jere H. Lipps, a geologist at the University of California at Davis, took the contrary view. In a paper in the journal *Science,* they argued that fossils in Galapagos sediments were no more than two million years old and that "all evolution of the islands' unique terrestrial biota occurred within the last three to four million years."

In their new paper, Dr. Christie and his colleagues noted: "Some have argued that the geological youth of the present islands, less than or equal to three million years, requires that all adaptive radiation has occurred within that period. Our geological observations and radiometric data indicate that there may have been islands present over the Galapagos hot spot for at least nine million years and probably much longer."

The geological evidence, Dr. Christie said, shows there is a "hot spot" in the Earth's mantle to which the Galapagos owe their creation. Molten lava from the hot spot burst out onto the floor of the Pacific Ocean, building up underwater volcanoes that eventually broke above the ocean's surface to form the Galapagos chain. The hot spot itself remains more or less stationary, but the crustal plate of the ocean floor has drifted eastward across it. The eruptions thus appear from the surface to move westward, with the westernmost islands being the newest; a half dozen volcanoes are still active on Isabela Island at the western end of the chain.

But at the eastern end, the oldest islands are no longer bolstered by fresh eruptions once they have moved past the hot spot. Erosion and the movement of the tectonic plates have combined to pull down and "drown" the oldest islands, presumably, the very islands on which iguanas first arrived from the mainland. Animals living on older islands would presumably have moved in a succession of migrations to newer islands in the chain.

A similar situation prevails in the Hawaiian Islands, where the crustal plate moves steadily across a volcanic hot spot, periodically creating new islands. But while most of the native animals and birds that evolved in the Hawaiian Islands were exterminated by early human settlers or by the animals the settlers introduced, the original Galapagos fauna have been somewhat better protected, and many unique species have survived.

Scientific understanding of the evolution of species began in the Galapagos in 1834, when the young Charles Darwin stepped ashore from the H.M.S. *Beagle* to collect samples. Astonished by the great variety of previously unknown birds, reptiles and mammals inhabiting the islands and their waters, he killed and preserved as many as he could find. Returning to England, Darwin showed his specimens to the ornithologist John Gould, who recognized them as a scientific treasure.

Among the birds, Mr. Gould discerned 13 different varieties of finches. All were obviously related to each other, but with markedly different beaks, some adapted to crushing nuts, others to cracking seeds or drilling insects from tree bark. As Darwin thought about the differences among the 13 finches, he realized that the variations in their beaks had developed from a single ancestral line to take advantage of different ecological niches. From this conclusion, the theory of evolution was eventually born.

Dr. Sarich said that his investigation of the proteins of Galapagos animals had not included the historic Darwin finches, but that in general, evolutionary paces for most animals were similar, except for one group of rapidly evolving rodents.

"I'm sure investigators will eventually get around to the finches and other species," Dr. Sarich said. "The Galapagos Islands are a never-ending source of fascinating scientific material."

—MALCOLM W. BROWNE, January 1992

How the Whale Lost Its Legs
And Returned to the Sea

AGES after some adventurous (or misadventurous) fishes left the sea and planted the flag of vertebrate animal life on land, their descendants had it both ways as amphibians and then completed the epic transition, evolving into terrestrial reptiles, mammals and birds. But something about the water must have kept beckoning, until a few irredentists among the mammals did eventually reclaim a place in the sea.

Most prominent of these mammals are the whales. Although they may swim the oceans with power and grace, these leviathans are more closely related to the camel and cow than any fish in their wake. Their anatomies retain vestiges of the four-legged land animals in their ancestry, the ones that began the bold return to the sea more than 50 million years ago.

As early as Aristotle, people recognized that more than size set whales apart from other marine life. Aristotle noted that whales as well as dolphins, the other members of the cetacean group, bear their young live, unlike fishes. This made them, by definition, mammals. (It is now known that some fishes do bear live young.) But even Darwin much later could not fathom the evolutionary steps by which some land mammals had become whales.

"It was a big evolutionary plunge, one that has eluded documentation for many decades," said Dr. Michael J. Novacek, a paleontologist who is provost of science at the American Museum of Natural History in New York City.

New fossil discoveries have now revealed several of the critical evolutionary steps in the earliest history of whales. Scientists have identified some intermediate species as land mammals that steadily changed physical form while adapting to swimming, diving, feeding and otherwise thriving in their new habitat. One surprise is that the transformation of four-legged land mammals into an animal completely adapted to marine

143

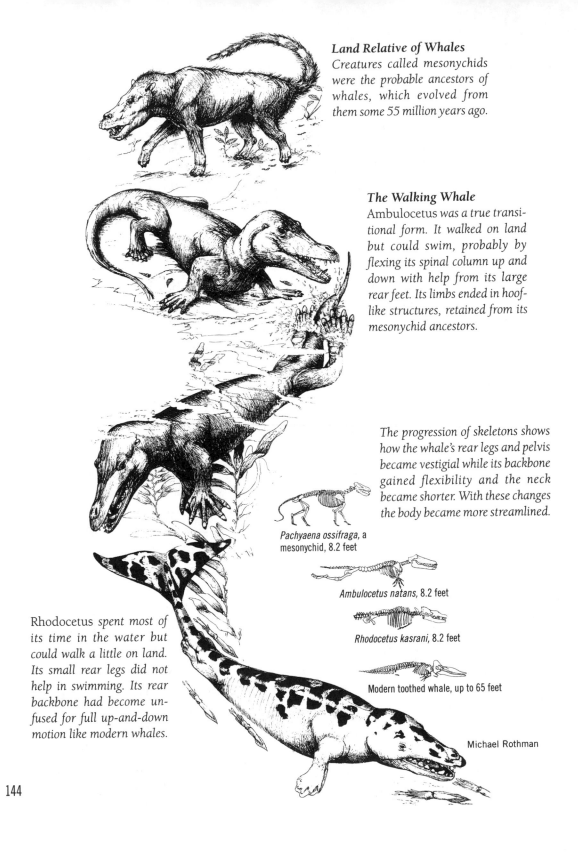

Land Relative of Whales

Creatures called mesonychids were the probable ancestors of whales, which evolved from them some 55 million years ago.

The Walking Whale

Ambulocetus *was a true transitional form. It walked on land but could swim, probably by flexing its spinal column up and down with help from its large rear feet. Its limbs ended in hoof-like structures, retained from its mesonychid ancestors.*

The progression of skeletons shows how the whale's rear legs and pelvis became vestigial while its backbone gained flexibility and the neck became shorter. With these changes the body became more streamlined.

Pachyaena ossifraga, a mesonychid, 8.2 feet

Ambulocetus natans, 8.2 feet

Rhodocetus kasrani, 8.2 feet

Modern toothed whale, up to 65 feet

Rhodocetus *spent most of its time in the water but could walk a little on land. Its small rear legs did not help in swimming. Its rear backbone had become unfused for full up-and-down motion like modern whales.*

Michael Rothman

life took only 10 million years—hardly any time at all in evolutionary terms.

"I am absolutely delighted to report that our usually recalcitrant fossil record has come through in exemplary fashion," Dr. Stephen Jay Gould, a Harvard University paleontologist and essayist, wrote in *Natural History* magazine. "The embarrassment of past absence has been replaced by a bounty of new evidence—and by the sweetest series of transitional fossils an evolutionist could ever hope to find."

One discovery, which Dr. Gould called "a remarkable smoking gun," introduced scientists to an animal that lived 50 million years ago. It was about the size of a modern sea lion, weighing 600 to 700 pounds and measuring about 10 feet from snout to tail. Judging by fossil remains, the animal was an amphibious species clearly intermediate between a terrestrial ancestor of whales and aquatic modern whales.

This fossil animal, named *Ambulocetus natans,* which means "swimming walking whale," was excavated from sediments of an ancient seabed in Pakistan. Dr. J. G. M. Thewissen, an anatomist and paleobiologist at Northeastern Ohio Universities College of Medicine in Rootstown, Ohio, reported the discovery in January.

Ambulocetus was just what scientists would have expected to find. The animal still had four limbs for walking on land, though probably with diminished agility. It could also hunt in the sea, probably swimming by kicking its big feet.

Another discovery in Pakistan has now advanced the transition story three or four million years and out into deeper waters. Scientists reported that they had found fossils of an animal that had obviously made a more resolute commitment to a life at sea.

Writing in the journal *Nature,* Dr. Philip D. Gingerich, a paleontologist at the University of Michigan at Ann Arbor, said the new species, *Rhodocetus kasrani,* was the earliest known transitional whale with an anatomy adapted for swimming like a whale. It had a more streamlined body and a fully flexible rear spinal column, which could have produced the motions for the powerful beat of a horizontal tail fluke that propels modern whales.

Whether *Rhodocetus* had indeed made this important advance cannot be determined until more complete tail fossils are uncovered. But the evidence, Dr. Gingerich and colleagues concluded, "shows that tail swimming evolved early in the history of cetaceans."

In a commentary accompanying the *Nature* report, Dr. Novacek said the *Ambulocetus* and *Rhodocetus* findings provided a "fascinating mosaic of early adaptive experiments in the transition from land to water."

Besides whales and dolphins, two other major types of mammals have made this dramatic transition: *Sirenia,* which includes dugongs and manatees, and the group composed of seals, sea lions and walruses.

The migration of whales to the sea is a classic example of transcendental evolutionary change, along with the first fish-like animals establishing their beachhead 370 million years ago, birds becoming flying animals and early human ancestors developing upright walking. But why would animals like whales, with all the apparatus for successful living on the land and ages of experience doing it, seek such a different habitat?

"The motive for any such transition is always opportunity," Dr. Gingerich said. "Look at what changed first, the teeth. What was initially important in the transition was food. These land animals were exploiting an opportunity to feed on fish in the sea."

Other changes over time involved improvements in the animals' maneuverability in their new environment. These included hearing systems evolved for underwater communications and navigation and completely new ways of locomotion, first as amphibians and then as fully developed marine animals.

For a long time, scientists had almost no direct evidence about the nature of the whale transition. The best they could do was make inferences based on a comparison of modern whales and fossils of archaic whales with the remains of their putative terrestrial relatives.

From this scientists concluded that whales were distantly related to ungulate mammals, a group whose modern members include camels, cows, pigs and deer. The link between other ungulates and whales is thought to be mesonychids, extinct four-legged mammals that sometimes feasted on fish at river edges.

Beginning in 1983, paleontologists have collected more revealing evidence of the transition. In that year, Dr. Gingerich reported the discovery of the oldest whale, which lived about 52 million years ago. A fossil skull was found in Pakistan in river sediments near an ancient sea. The animal was given the name *Pakicetus.*

Although they had nothing to ponder but this skull, scientists could see that *Pakicetus* had teeth resembling those of mesonychids, but it was well adapted to feeding on fish in surface waters of shallow seas. Other parts of the skull were becoming whale-like, but it lacked the auditory equipment for a fully marine existence. Another fossil specimen from about the same time, *Indocetus ramani,* probably led the same kind of life, entering the sea to feed on fish, but returning to land to rest and raise its young.

As Dr. Gould noted in his review of recent whale research, the next important discovery was the first complete hind limbs of a fossil whale, *Basilosaurus isis,* which lived 5 million to 10 million years after *Pakicetus.* The skeleton, found in Egypt, was described in 1990 by Dr. Gingerich and a team of Michigan and Duke University scientists.

Since the hind limbs were a mere two feet long and the whole body was 50 feet long, the discoverers concluded that the legs could not have supported the body on land or assisted in swimming. This fossil whale had passed the point of intermediacy, and so the search continued.

If the matter of locomotion is the functional test of intermediacy, as Dr. Gould wrote, Dr. Thewissen's *Ambulocetus* came closest to the long-sought breakthrough in early whale evolution. Its toes are even terminated by hooves as in mesonychids and other ungulates, those earlier ancestors. The skeleton also revealed that the animal had stubby forelimbs with hands that splayed outward like the flippers on a sea lion, while the rear legs were still large and powerful, with possibly webbed feet. It had not yet evolved a tail fluke, but its spine seemed to be flexible enough to allow the undulations associated with whale propulsion.

In this intermediate phase, the up-and-down undulations probably operated in concert with the vigorous paddling of *Ambulocetus's* large feet. In modern whales, they contribute mightily to the propulsive beat of the tail fluke.

On the basis of this evidence, Dr. Thewissen concluded, "*Ambulocetus* represents a critical intermediate between land mammals and marine cetaceans."

Whales in their present form began appearing about 30 million years ago. Their flippers are what remains of the forelimbs of their terrestrial past.

The only hints of the former hind limbs are the vestiges of a pelvis and femur, the upper-leg bone, embedded in the body wall.

Reflecting on the recent succession of discoveries, Dr. Novacek said, "This expanding fossil casebook on the origins of whales is one of the triumphs of modern vertebrate paleontology."

—JOHN NOBLE WILFORD, May 1994

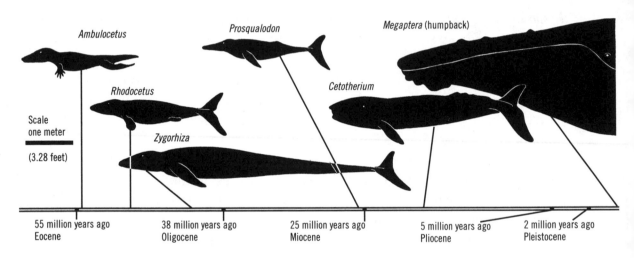

Michael Rothman

Treasure Trove of Fossils
Turns Up in the Gobi

NO PALEONTOLOGIST had ever investigated this part of the Gobi Desert, some low hummocks of 80-million-year-old reddish rock in a remote basin in southwestern Mongolia. Not Roy Chapman Andrews, the adventurous American explorer of the 1920's. Not the Russian and Polish scientists who reconnoitered there more than two decades ago. And last summer's American-Mongolian expedition might have passed through without striking real pay dirt, if it had not been for what happened to one of the trucks.

As Dr. Michael J. Novacek, the expedition leader, told the story, the caravan of dust-coated trucks had traveled some 250 miles southwest of the town of Dalan Dzadgad through mountains and over sand dunes, across some of the bleakest landscape in the Gobi. About 30 miles north of the forlorn village of Dauz, the party approached the basin, which did not look like much from a distance.

"We got there after some struggle with the gasoline truck," Dr. Novacek said blandly in a recent interview. Whereupon another member of the science team who was present, Dr. James M. Clark, interrupted to volunteer a revealing amplification. "Yes, we made camp where the gas truck got stuck in the sand," he said.

Right there, on the ground by their trucks and everywhere they walked, the paleontologists began picking up fossil bones lying on the surface. In the next three hours, they found 30 skulls of lizards and tiny mammals, all remarkably well preserved. In contrast, in five visits in the 1920's, the Andrews expeditions collected only four mammal skulls.

In 10 days at the basin, called Ukhaa Tolgod by the Mongolians and nicknamed Xanadu by the wide-eyed American fossil hunters, the expedition collected the skulls and skeletons of 13 theropod dinosaurs (flesh eaters

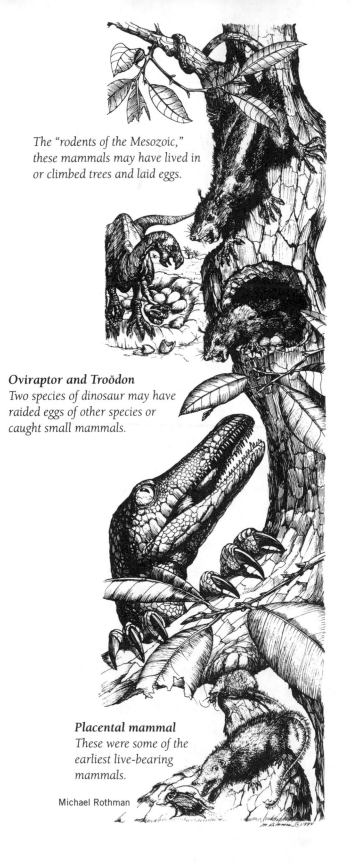

The "rodents of the Mesozoic," these mammals may have lived in or climbed trees and laid eggs.

Oviraptor and Troödon
Two species of dinosaur may have raided eggs of other species or caught small mammals.

Placental mammal
These were some of the earliest live-bearing mammals.

Michael Rothman

150

that walked on their hind limbs), 175 lizards and, perhaps most important, 147 mammals. They found several dinosaur "nurseries" with clutches of fossilized eggs. As many as 100 other dinosaurs were identified and left to be picked up another year.

It was the greatest triumph so far for the expedition of the American Museum of Natural History in New York City and the Mongolian Academy of Sciences. Last summer, in their third season of fieldwork, the scientists had identified the most concentrated repositories of some of the most varied and abundant fossils from the geological era known as the Cretaceous, which extended from 145 million years ago to 65 million years ago, the time of mass extinctions in which the dinosaurs vanished.

"What we found—and there is no doubt we were here first—is perhaps the richest little pocket of bones in the Cretaceous Gobi, or the Cretaceous anywhere for that matter," said Dr. Novacek, vice president and provost of science at the American Museum.

This time no one amended the leader's words.

Even Dr. Malcolm C. McKenna, a team member and a curator of paleontology at the museum who is not usually given to verbal exuberance, was excited about the discoveries at such a promising new site. "There's no time like the first time in paleontology," he remarked.

Other expedition scientists, including Dr. Demberlyin Dashzeveg of the Mongolian Academy, who is visiting the museum, also described their good fortune in interviews. Although they and laboratory assistants are still cleaning the fossils and examining the more intriguing ones in greater detail, their rich haul is expected to include several new species of dinosaurs and early mammals and more evidence linking dinosaurs to the origin of birds.

The discovery, though accidental, was not completely unforeseen. On the advice of Dr. Dashzeveg, the party had headed for this area at the foot of the Nemegt Mountains, a place he had heard might have exposed sediments from the late Cretaceous period. But the stuck truck put the scientists at a fossil lode they might have overlooked.

The basin, about one mile square, was more hospitable to life 80 million years ago, with a semi-arid climate and some standing water. But wind-blown sands must have buried the dinosaurs, lizards and tiny mammals soon after they died, assuring them a certain immortality as fossils.

On the second day at the site, as Dr. Novacek was walking in the flats near a couple of buttes dubbed Camel's Hump, he came across six dinosaurs, ankylosaurs with spiky tails, lying exposed on the surface. Over by a ledge, he saw another skull embedded in the rock, then most of the rest of its skeleton. But this one was no ankylosaur.

Dr. Mark A. Norell, one of the expedition's dinosaur experts, identified it as a *Troödon* and possibly one of the major finds. These were relatively small, agile dinosaurs, about six feet long, whose fossil remains are rare.

"This is a very, very complete specimen," Dr. Norell said after examining it in the laboratory. The distinctive shapes and varied sizes of the teeth indicate that it probably represents a heretofore undiscovered species of this type of dinosaur, whose place in evolution is thought by some scientists to have a bearing on the origin of birds. Some *Troödon* fossils have been mistaken for bird bones.

In a previous season in the Gobi, the paleontologists had dug up fossils of an animal about the size of a turkey that they contend was a flightless bird, a transitional figure between some carnivorous dinosaurs and modern birds. They named it *Mononychus olecranus*. At the new site, they found another 10 specimens of *Mononychus*, including a much more complete skull than previously known.

Dr. Norell said the new findings appeared to strengthen their hypothesis that *Mononychus* was "clearly a basal avarian" and more closely related to birds than the 150-million-year-old *Archaeopteryx*. The latter was a bird-like dinosaur with feathers, which has been considered a critical link in the evolution of reptiles to birds.

Another important dinosaur discovery was the complete skeleton of an *Oviraptor*, about five feet long. The Andrews expedition identified the first one at Bain Dzak, spectacular fossil beds that the explorer celebrated as Flaming Cliffs. The new find should give paleontologists their first complete description of the animal's skull.

Although the Gobi has been known primarily for its dinosaur fossils, the multitude of small mammal bones uncovered at the new site could be the expedition's biggest prize. Cretaceous mammal fossils are virtually absent outside Mongolia. Many of the 147 specimens were found encased in sandstone nodules lying on the ground, where they had recently eroded out of the hillside.

The earliest mammals were shrew-like creatures that appeared some 200 million years ago. These and later small mammals lived in the shadow of the giant dinosaurs, occupying the ecological niche that would eventually be taken over by rodents. Among the more common mammals of the Cretaceous were a group known as multituberculates, probably egg-laying animals about the size of mice and rats that ate seeds and nuts and are long extinct.

At Ukhaa Tolgod, the paleontologists picked up an abundance of these especially well-preserved "multi" skulls and skeletons, as well as a few specimens of placental mammals, whose young are born live and fairly well developed. The placental skulls range from less than an inch to two inches long.

"The late Cretaceous was a critical time for mammalian evolution, a time of the early expansion of placental mammals," said Dr. McKenna, a specialist in early mammals. "This may give us the first really good anatomy of placental mammals that far back."

The expedition's previous discoveries of mammal fossils in the Gobi and the collections of Russian and Polish paleontologists in earlier decades provide most of the important evidence for the evolution of mammals in the late Cretaceous.

"To put these discoveries in perspective," Dr. Novacek said of the new site, "consider that the total accumulation of Cretaceous mammal skulls collected over 70 years all over the Gobi amounts to probably something less than 100 skulls. It is extraordinary to consider a 10-day haul in one square mile of nearly 150 skulls."

Seventy years ago, in the flush of discoveries that first revealed the Gobi as a vast graveyard of late Cretaceous life, Roy Chapman Andrews of the American Museum exclaimed, "The stuff is here!" And he had never visited Ukhaa Tolgod, the place his successors call Xanadu. To fossil hunters the site is their idea of the "stately pleasure-dome" decreed by the great Mongol Kubla Khan, by way of the fevered imagination of Coleridge.

—JOHN NOBLE WILFORD, April 1994

Expedition to Far New Jersey
Finds Trove of Amber Fossils

AN AMERICAN MUSEUM OF NATURAL HISTORY expedition—to New Jersey—has uncovered one of the richest deposits of amber ever found, with fossils of 100 previously unknown species of insects and plants trapped in the ancient fossilized tree sap.

The fossils include a tiny bouquet of miniature flowers from an oak tree of 90 million years ago; the world's oldest mosquito fossil, with mouthparts tough enough to feed on dinosaurs; the oldest moth in amber, with mouthparts suggesting it was in transition from a biting insect to one that fed on the nectar of flowers, and the oldest biting blackfly. The last is the only such insect known from the Cretaceous period, and may have tormented duckbills and other dinosaurs along with its colleague in amber, the mosquito.

Among the other finds from the New Jersey complex of sites are the oldest mushroom, the oldest bee and a feather that is the oldest record of a terrestrial bird in North America.

Dr. David A. Grimaldi, curator and chairman of the entomology department at the American Museum of Natural History, who led the expedition to the secret New Jersey site, said the previously undescribed species, all extinct, were found in 80 pounds of amber drawn out of deep mud.

The ambers came from sites in central New Jersey where the clay is especially deep and rich, containing streaks of peaty black material that are the remains of plants and other organic material. It is in these rich black stripes that the amber was found.

What is most interesting to scientists is that the site has amber dating back 90 million to 94 million years. This means that all the amber-preserved species came from the age of the dinosaurs and from the era when flowers first began to proliferate. At the time, insects were beginning to use flowers as food, and flowers found the insects useful in carrying pollen from flower to flower.

An article describing the world's oldest preserved flowers, written by Dr. Grimaldi and his colleagues Kevin Nixon and William Crepet of Cornell University, is published in *The American Journal of Botany*. It notes that the three flowers in the little bouquet are the only known flowers preserved from the Cretaceous period, which ended 65 million years ago.

One reason the fossil flowers are important is that the flowering plants that arose during the Cretaceous eventually took over the Earth's continents, dominating the landscapes in both their numbers and importance to ecosystems.

Until now, the study of plants from the Cretaceous has depended solely on impressions of flowers and pollen. Curiously, the flowers and some of the other fossils found at the New Jersey site are miniatures—the flowers and their stem together are no more than half an inch long.

The new finds also raise the problem of what to do with the specimens. People have valued and studied creatures in amber—ants, bees, scorpions, lizards, frogs—for several thousand years. In the 19th century, biologists observed the insects without breaking open the amber, and classified them by physical features.

In the last few years, scientific interest has grown in the DNA locked inside the creatures inside the amber. There is a great temptation to compare it with the DNA of similar modern species. Dr. George Poinar, now at Oregon State University, one of the scientists credited with the discovery that bits of intact DNA exist in fossils inside amber, said that the material inside the insects "is the best-preserved protein on the face of Earth." He said, "Not only does the amber draw out the water to dehydrate the specimen, but the terpenes in the resin act as a fixative."

Inside the insects the preservation was great enough to have kept even muscle tissue intact in a 125-million-year-old Lebanese weevil that Dr. Poinar studied. To date, scientists have successfully extracted bits of DNA from half a dozen amber drops.

But should the amber be cracked open? A dispute has arisen in biology about how to handle amber specimens—whether to open them to get at their DNA, how to open them and whether there should be some rules guiding new expeditions and the use of existing collections.

There are tens of thousands of pieces of amber with fossil insects and other items in collections around the world, and most of them are common varieties of ants and flies. The rarest pieces contain lizards, frogs or the hairs of mammals. The color of amber ranges from the rare blue and green pieces to the more common yellow, orange and red. They can be as small as a tear or as large as a softball.

Some several thousand of the specimens contain the only examples of now-extinct species and so are true biological treasures.

Dr. Grimaldi urges caution, saying, "The possibility of studying DNA in amber fossils is exciting but also represents a serious problem." Because all creatures on Earth are classified according to their physical features, and it is this that scientists use to study how evolution has created or shaped the entire history of life, Dr. Grimaldi said he believes that no specimens in amber should be tampered with unless they are quite common and others of the same species, era and location can be found to replace them.

Further, because the art of extracting DNA from amber fossils is just being developed, it is likely that for every success, three or four specimens will be damaged with no useful information gained.

Dr. Grimaldi has criticized the extraction work of scientists like Dr. Poinar, pointing out that the weevil from Lebanon was probably the only one of its kind in the world.

Dr. Grimaldi said his conservative approach is becoming more common in science, from archaeology to paleontology, in which large parts of discovery sites are left unexcavated, so as not to destroy all evidence that might be useful to future scientists. The idea is that techniques of the future will be able to gain much more information from the same material, and random digging now might destroy it.

All amber fossils, Dr. Grimaldi says, should be coded, as is now being done at the American Museum of Natural History, according to rarity, and only those with more than two or three duplicates should be subjected to DNA extraction.

Dr. Poinar, for his part, agreed that very little damage can be tolerated to rare amber fossils. But he said he and others have developed techniques to enter the fossil specimen through a tiny hole, use a needle to extract tissue with DNA and then reseal the hole.

"This work will be going ahead, and people will be opening amber anyway, whatever we say, so we must develop ways to prevent or minimize damage to the specimens," Dr. Poinar said. The fossil insects can be kept almost perfectly intact, and besides, the marring of the surface of the insect with needle extraction causes minor damage only to one side of the specimen. The other remains untouched.

No good, says Dr. Grimaldi. "Not even the most minimal kind of damage is acceptable," he said. In one article on amber, he quoted from the poem "Hesperides," by Alexander Pope:

> I saw a fly within a bead
> Of amber clearly buried;
> The urn was little, but the room
> More rich than Cleopatra's tomb.

Scientists may be willing to plunder the tomb of Cleopatra, but Dr. Grimaldi argues they should refrain from the plunder of amber tombs. "An obsession with technological feats," Dr. Grimaldi wrote, "makes us into mere tinkerers, and distances us from social and scientific ethics."

—PHILIP J. HILTS, January 1996

Reservoir of Living Fossils
May Lie Below Ice

PROBING THE GLACIAL DEPTHS of Antarctica with radio and seismic waves, scientists have mapped the locations of more than 70 lakes that lie buried under thousands of feet of ice. The largest of these, some 140 miles long and 30 miles wide, is a body of apparently fresh water more than two miles beneath the ice, next to Russia's Vostok research base in East Antarctica.

The existence of the subglacial Lake Vostok has been known for two decades, but only recently have its vast dimensions been determined and its potential importance to science been recognized. Some researchers suspect that the lake's pristine waters contain living "fossil" microbes, bacteria and viruses essentially unchanged over as much as a million years.

But as eager as scientists are to sample these organisms, they are at a loss to figure out how to do it without contaminating the lake in the process. An international group of scientists, meeting in 1995 at Cambridge University in England, agreed on the value of sampling the lake, but stressed the need to conduct a thorough study of how to drill to the water and prepare an environmental impact assessment of the project. One possibility, the scientists said, would be to test sampling techniques at a smaller lake before starting work at Lake Vostok.

At a conference in 1996 in Utrecht, in the Netherlands, representatives of the nations involved in Antarctic research endorsed the Cambridge group's recommendations and adopted a resolution urging Russia to halt its planned ice coring at the Vostok station at a safe distance—no closer than 75 to 150 feet—above the subglacial lake to avoid any risk of polluting it. Participants at the conference said the Russians and their French and American partners in the drilling project were expected to take these precautions when work resumes.

In a telephone interview, Dr. David L. Clark, a geologist at the University of Wisconsin at Madison, who is chairman of the Polar Research Board of the National Academy of Sciences, said, "The Russians are keenly aware of the fact that this is a unique fossil lake."

Dr. Julie M. Palais, director of the Antarctic Glaciology Program at the National Science Foundation, said there were no plans so far for sampling the lake's waters, but proposals from scientists could be expected soon. The foundation coordinates and finances American research in Antarctica.

The Russian drilling program is the most ambitious ever undertaken in Antarctica, collecting ice samples to study climate change over the ages. Russian scientists reported at Utrecht that over 20 years they had reached depths of more than two miles. During the 1995–96 research season, they penetrated within 1,500 feet of the lake. French scientists who analyzed the deepest ice samples said they were as old as 420,000 years. This has led to estimates that Lake Vostok has been sealed under the ice for at least 500,000 years, perhaps as much as a million years.

Seismic studies by the Russians indicate that the lake water could be more than 2,000 feet deep, with bottom sediments about 500 feet deep at one end. The Russians said the lake basin could be a rift valley, a deep split in the Earth's crust.

Dr. Michael Bender, an oceanographer at the University of Rhode Island who specializes in Antarctic research, said that heat from the Earth's interior could be keeping the deep waters from freezing. Although radio-echo sounding sensitive to the slightest changes in density suggests that the water is fresh, some scientists raise the possibility that it could be ancient seawater that was trapped long ago in the basin.

The charting of subglacial lakes has mainly been done by British scientists flying radio-sounding instruments on United States Navy planes. With one third of the continent surveyed, about 70 lakes have been detected, most of them at least 9,000 feet beneath the surface and a few miles wide and up to 40 miles long.

—JOHN NOBLE WILFORD, May 1996

Clash on Fossil Sales Shadows a Trade Fair

THERE WAS SOMETHING for everyone: $5 fossil sea urchins for customers on tight budgets, museum-quality dinosaur skeletons at prices up to $1.1 million and hundreds of thousands of other fossils of all types and values.

But as crowds packed the Tucson Minerals and Fossils Show to ogle the paleontological treasure on sale, Federal officials were looking for new ways to crack down on fossil trading.

Thousands of commercial dealers, collectors, museum agents and fossil buffs from many nations filled the hotels and motels of Tucson, where the annual show spread out over many square miles.

Hundreds of hotel rooms, banquet halls and meeting rooms were converted into private museums during the week for an event that has become the largest trade fair for fossil dealers throughout the world. Eye-popping fossils from China, Germany, England, Spain, Morocco, Lebanon and many other countries joined a wealth of fossils from America's Badlands and bluffs in the big sale.

But it was a trade fair steeped in dispute, bitterness and litigation, with commercial dealers claiming a legitimate right to hunt, collect and sell fossils, while government agencies and some paleontologists strive to halt the trade.

Commenting on the current boom in fossil sales and prices, Dr. William S. Clemens, a professor of paleontology at the University of California at Berkeley, and the president of the Society of Vertebrate Paleontology, said: "I've been visiting the exhibits at the Tucson show this year to get a feeling for the trade, and some of the things I saw made me sick. I saw some exhibits marked with numbers similar to those used by museums, and I couldn't help wondering whether these specimens had been looted from museums. I saw a rare fossil amphibian from Russia on sale, accompanied by a certificate from Russia's Paleontological Institute allowing export of this treasure. The Russians must certainly be hard up to let things like that go."

Although many a deal was done, the trading this year took place under a legal cloud that could send some prominent fossil dealers to jail for terms of hundreds of years. With the Justice Department, the Federal Bureau of Investigation, the Bureau of Land Management, the National Park Service and other Federal agencies breathing down the necks of America's fossil collectors, dealers here were taking pains to avoid even the hint of wrongdoing.

Several criminal cases are pending against fossil dealers, and the Justice Department is seeking fresh evidence for use in prosecutions. Meanwhile, fossil dealers who gathered for the Tucson show, many of whom described recent government initiatives as a "witch hunt," were organizing to defend themselves by legal action, perhaps aided by a Washington-based lobby. The charges and implications levied against the dinosaur dealers are serious.

Dr. Dan Chure, a National Park Service paleontologist at Dinosaur National Monument in Utah, said in an interview that the Colombian drug cartels might have become interested in acquiring valuable stolen fossils that disappear from museums or private estates. John Kundts, a spokesman for the FBI in Washington, said that the prices of fossils, particularly those of dinosaurs, had become high enough to attract criminal notice.

Fossil dealers here angrily deny implications that their businesses have any ties to organized crime, and have accused government agencies of slandering them.

Federal agencies, spurred to action by opponents of commercial fossil collecting, have undertaken a costly campaign against some fossil dealers who sell their wares to museums, corporations and private collectors in many countries. Targets of this campaign could eventually include educational and scientific institutions that have acquired fossils from dealers, and officials of museums in several countries are uneasy.

A Federal indictment issued last November named the Smithsonian Institution in Washington and the Field Museum of Natural History in Chicago as buyers of fossils excavated illegally from Federal lands. Karen E. Schreier, United States Attorney for South Dakota, where the indictment was issued, said in an interview that she did not intend to prosecute either museum because they had cooperated with investigators.

But dealers here said that museums in Europe that had lent many fossils to American institutions were worried that future action by the Federal Government was unpredictable, and that Federal officials might even seize collections on loan from abroad. Several European museums are therefore recalling their specimens from American museums.

Some American curators are also uneasy about a questionnaire that will soon be circulated by the National Park Service requesting an accounting by museums of all archaeological and scientific materials in their collections "obtained through relationships with Federal agencies such as permits and contracts." Although response to the questionnaire is voluntary, some museum officials fear the results of the survey could become the basis of Federal legal actions against some institutions.

The large investment the government has made in prosecuting fossil dealers has awed collectors, scientists and museum officials.

The storm center of the dispute is the Black Hills Institute of Geological Research, Inc., of Hill City, South Dakota, and its officers, who, under the Federal indictment issued last November, could each face long prison sentences and fines for illegally collecting and selling fossils excavated from Federal land. The case involves the bones of a tyrannosaur, nicknamed Sue.

According to Gary G. Colbath, one of the defense lawyers involved, the Justice Department has already spent $4.5 million preparing its complex case against the Black Hills Institute and its president, Peter Larson. The preparation has included repeated raids on the Black Hills workshop, a fact-finding mission to Japan, where museums have bought fossils from the company, and the deployment of scores of Federal agents and National Guardsmen.

Professional paleontologists are divided in their opinions about commercial fossil dealing in general and the Larson case in particular.

Dr. Robert M. Hunt, a paleontologist at the University of Nebraska in Lincoln, has spearheaded a drive to restrict the activities of commercial collectors, and has expressed delight at Mr. Larson's indictment.

"Unfortunately," he said in an interview, "the abuse of America's fossil resources by these dealers is getting worse than ever. I can only hope that a public outcry against this assault on the resources of science and the American people will help to halt trafficking in priceless fossils."

A milder view was expressed by Dr. Clemens. "It comes down to a sense of responsibility," he said. "Many collectors have worked closely with scientists and museums, to their mutual benefit. Large museums have their own resources and staffs, but what's a small museum to do if it wants to acquire some good specimens? I think we must encourage large institutions to reach out and help smaller museums, and that has begun to happen."

Dr. John Maisey, an English ichthyologist and paleontologist at New York City's Museum of Natural History, declared: "This whole campaign by the Federal cops against the fossil business is ridiculous, considering all the murderers out there that remain to be caught. If they really want to stop the under-the-counter fossil trade, they should legalize it on Federal land, and regulate the practice, as governments in Europe do. This assures scientists and museums of access to important fossils, but allows a legitimate trade to exist."

Many of the items displayed at this year's show represent close collaboration between commercial dealers and scientists.

In the Quality Inn lobby in Tucson, Michael Triebold, a dealer from Valley City, North Dakota, set up a 20-foot-long, freestanding articulated fossil of a *Xiphactinus audax,* a meat-eating fish with huge jaws and teeth, that lived at the time of the dinosaurs. The fossil is shown being attacked by a fossil mosasaur, a large reptile predator of the shallow Cretaceous seas. Both are casts of the original fossils, which were collected in Kansas.

"You don't have to hunt on Federal lands to find good fossils," Mr. Triebold said. "I enter into a partnership with landowners, and if we find and excavate fossils, the landowners share in the profits of the sales."

In some regions, plots of federally owned land are interspersed with private land, and since there are no markers or fences, it is difficult to know which is which.

"I take no chances," Mr. Triebold said. "I use a Global Positioning System satellite receiver to locate sites of interest. But a GPS measurement may be off by several hundred yards, and when there is any doubt, I hire a surveyor to locate the boundary between Federal and private land."

The treasures found by Mr. Triebold's company are rich beyond the dreams of many museums. In his little showroom here, he displayed the leg bone of a triceratops dinosaur collected in South Dakota, which he calls a "Rex Biscuit." It is perforated by the dagger-like teeth of a tyrannosaur's jaw, and it is deeply gouged by the fangs of lesser carnivores that scavenged the leavings of the tyrannosaur's meal 65 million years ago. Mr. Triebold sells casts of this remarkable specimen for $850 each. Although he prefers to sell casts, he would sell the leg bone, incorporated into a complete triceratops skeleton, for $500,000.

Most of the deals here involve the sale of casts rather than original fossils, and dealers say that some 90 percent of the customers are museums. But for institutions or people with deep purses, some original fossils are available that would make any museum curator drool.

Geological Enterprises, Inc., of Ardmore, Oklahoma, for example, is now cleaning and preparing an extremely rare fossil of an *Acrocanthosaurus atokensis,* a member of the carnivorous allosaur family that grew to enormous size in the early Cretaceous period. With a skull five feet seven inches long, this is the second largest carnivorous dinosaur ever found, following the tyrannosaur itself.

The company hopes to begin selling casts of the completed skull soon, but it would sell the entire fossil, prepared and mounted, if a suitable buyer is found. "The price is $1.1 million, take it or leave it," said Leon Theisen, whose South Dakota shop is preparing the fossil. "Naturally, we would try to sell it only to a museum or scientific institution."

Paleontologists are sharply critical of some fossil dealers who cut up dinosaur fossils to make decorative objects or jewelry, but some of these dealers defend the practice by saying they only use bone that has been broken up already and spoiled for scientific study.

At one of the stands here, a company in Moab, Utah, The Look of the Past, offered earrings and pendants made of dinosaur bone that had mineralized into agate. But along with the jewelry, the company displayed an allosaur leg bone reassembled from its broken parts and then sliced in two with a diamond saw. The exposed stone surfaces were highly polished, revealing a blazing pattern of colored agate that highlights the cell structure of the big dinosaur's bone tissue and marrow. Its price was $8,000.

But despite the high prices of some fossils, "none of us get rich, and you won't find any fancy cars around here," said Dr. Raymond J. Boyce, a retired South Dakota urologist who has collected fossils all his life. Dr. Boyce's son Japheth now heads the fossil dealership he founded in Rapid City, South Dakota. "The money people pay is not for the fossils, which cost nothing," he said. "It's for the months of hard searching and the study and craftsmanship needed to separate these fossils from the matrix rock and prepare them properly and look after them scientifically—that's what people pay for.

"The government called me before the grand jury to testify, and maybe it's getting ready to put me in jail, too," Dr. Boyce said. "Here we've been working as partners with our Indian neighbors for as long as anyone can remember, finding fossils, enjoying them, showing them and giving them to museums and sometimes selling them. I bought fossils from Ed Two Bulls, Sr., himself, a descendant of Sitting Bull. And now the Federal Government wants to jail us all. I tell you, this country is changing in ways I hate to see."

His son Japheth nodded grimly. "We don't have money to do much yet, but we're trying to organize a new group, the Earth Sciences Trade Association, to fight back and sue the people who slander us. We're not dead yet."

—MALCOLM W. BROWNE, February 1994

"Dwarf" Mammoths
May Have Put Off Demise

Woolly mammoth
—10.5 feet tall

Dwarf mammoth—5.9 feet tall

Michael Rothman

Teeth and bone fragments found on Wrangel Island in the Siberian Arctic suggest that a dwarf mammoth (shown here with full-size woolly mammoth) survived there for 6,000 years longer than any other mammoth species. Bones are 4,000 to 7,000 years old.

THREE RUSSIAN SCIENTISTS have turned up evidence that a race of "dwarf" mammoths, cut off from the rest of the world on a remote Arctic island, survived the extinction of all other mammoths by up to 6,000 years, living on into comparatively recent times.

The scientists' conclusion came after the discovery of 29 complete but very small mammoth teeth, as well as many tooth and bone fragments, on

166

Wrangel Island, a Russian possession in the Arctic. By analysis of proportions of carbon isotopes in the teeth, the scientists determined that the teeth were 4,000 to 7,000 years old, which is far younger than any previously known mammoth fossils.

Paleontologists previously believed that all mammoths were extinct by 9,500 years ago and that mammoths had disappeared from most of their ranges in Europe, North America and Asia by 12,000 years ago.

The mammoth teeth from Wrangel Island are up to 25 percent smaller than those of ordinary woolly mammoths, whose known fossils are all much older. The teeth were those of adult mammoths, which can be easily distinguished from those of juveniles by their cellular structure.

The scientists inferred from the dwarf teeth that these animals were more than 25 percent smaller than the full-size woolly mammoths; in other dwarf forms of animals, the reduction of body size is generally greater than the reduction of tooth size, and the Russian group assumes that the same is true of mammoths.

The discovery is reported in the journal *Nature* by Dr. S. L. Vartanyan of the Wrangel Island State Reserve, Dr. V. E. Garutt of the Zoological Institute of the Russian Academy of Sciences and Dr. Andrei V. Sher of the Severtsov Institute of Animal Morphology and Ecology in Moscow.

In an accompanying comment, Dr. Adrian M. Lister, a biologist at University College, London, estimated that the dwarfs would have stood about six feet and weighed about 2.4 tons. The full-size woolly mammoth of the species *Mammuthus primigenius,* the same type as that found, stood 10.5 feet and weighed 6.6 tons.

The explanation for the surprising survival of these dwarf mammoths is certain to be strongly debated among paleontologists.

Until about 16,000 years ago, the land that is now Wrangel Island was part of a mainland region known as Beringida, where animals could roam a vast region that included Arctic eastern Siberia, Alaska and much of what is today the shallow Arctic Sea shelf.

But by 12,000 years ago the land bridge between the mainland and Wrangel Island was cut off by rising seas, and it was then that the evolutionary process that produced the dwarf mammoths probably began.

Scientists have long debated whether environmental changes or human hunting drove mammoths and many other large mammals to extinction at the end of the Pleistocene epoch, and the Wrangel Island discovery is certain to fuel this debate, Dr. Lister said.

One explanation could be that the dwarfs survived on Wrangel Island because human hunters had not yet reached the island. Once people arrived, according to this theory, the extinction of the dwarf mammoths was assured, thus wiping out the last of a great line and leaving elephants as their only surviving cousins.

The Russian group discounts this idea, saying there is no evidence of human hunting connected with the Wrangel Island fossils. Instead, the group prefers another explanation: that after Wrangel Island was cut off from the mainland, some combination of local terrain and climate on the island preserved communities of plants that had formerly flourished on the Siberian steppes but were dying out there.

This steppe habitat may have been essential to mammoths, and only where it survived for a time were mammoths able to keep going. Even today, Wrangel Island retains a greater diversity of herb species than other Arctic islands.

But Dr. Lister points out that Eskimos are known to have inhabited Wrangel Island 3,000 years ago and to have hunted on Zhokov Island, also in the Arctic Sea, up to 8,000 years ago. The dates could well accommodate the arrival of early hunters on Wrangel while the mammoths were still around.

The Russian group says the rapid evolution of normal-size species into dwarfs in isolated island environments is known among other species.

—MALCOLM W. BROWNE, March 1993

Horses, Mollusks and the Evolution of Bigness

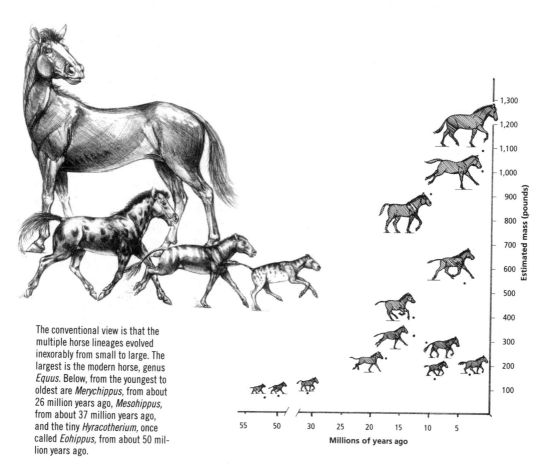

The conventional view is that the multiple horse lineages evolved inexorably from small to large. The largest is the modern horse, genus *Equus.* Below, from the youngest to oldest are *Merychippus,* from about 26 million years ago, *Mesohippus,* from about 37 million years ago, and the tiny *Hyracotherium,* once called *Eohippus,* from about 50 million years ago.

Millions of years ago

Estimated mass (pounds)

In fact, once diversification of horses began, about 30 million years ago, multiple lineages produced animals with a wide range of masses and statures, and they persisted over most of horse history. Until recently, small species, medium-size species and larger ones existed simultaneously, It is an anomaly that the only surviving species happens to be the largest.

169

Michael Rothman

PALEONTOLOGISTS in the 19th century kept digging up fossils of early horses no bigger than a diminutive mouse deer. Over the expanse of evolutionary time, they came to see, the horse lineage showed tendencies to increase in body size, leading to the sole surviving genus *Equus,* which includes such imposing animals as thoroughbreds and Clydesdales.

Other discoveries seemed to confirm a trend toward bigness. Many of the giant dinosaurs were coming to light at this same time. Even the tiny Foraminifera, single-celled plankton so common in the marine fossil record, appeared to follow a pattern of lineages starting with small founding members and then evolving larger body sizes.

With such observations in mind, the American paleontologist Edward Drinker Cope formulated in 1871 a "law" of biology: The body size of organisms in a particular evolutionary lineage, whether horses or mollusks or plankton, tends to increase over the long run. This principle, known as Cope's Rule, ascended to conventional wisdom in biology. It is cited in textbooks to this day.

For all these years, however, scientists had never bothered to put Cope's Rule to a rigorous test. A few apparent exceptions had been exposed, and over the last decade, some prominent scientists questioned the rule's general validity. Now a University of Chicago paleontologist has marshaled the most comprehensive evidence yet to challenge Cope's Rule.

Completing a comprehensive study of mollusks over a 16-million-year period, Dr. David Jablonski concluded that there is no more tendency for organisms to become bigger as they evolve than there is for them to become smaller. Even in the lineages showing size increase for the largest species, he noted, the smallest species often decrease in size over the same interval. The result is an expansion of the overall range of variation, both large and small, and not, as Cope had surmised, a directed trend toward increasing general size.

The study is based on an analysis of fossil clams, snails, whelks, scallops and other mollusks that lived on the coastal plain from New Jersey to Texas in the last 16 million years of the Cretaceous geological period, which ended with the mass extinction 65 million years ago that killed off the dinosaurs. Dr. Jablonski spent the last decade, with a caliper always at the ready, examining museum fossil collections of 191 evolutionary lineages, including precise measurements of more than 1,000 species.

"This is the first time anyone has taken a quantitative look at a large-enough data base to really draw a general conclusion," Dr. Jablonski said.

Dr. Douglas Erwin, a paleontologist at the Smithsonian Institution in Washington, praised the care and methods of Dr. Jablonski's study, saying it "set the standard" for future research into broad trends of evolutionary biology. He suggested that there were probably "many other trends that paleontologists think they see in life, think are true, but often don't have rigorous tests to prove them."

Dr. Daniel McShea, a Duke University paleontologist, said that similar studies should be conducted on a wider variety of life in different time intervals. "That would really test Cope's Rule," he said. But he agreed that the new findings showed that the rule is by no means an invariable law of nature.

The results of Dr. Jablonski's research were published in the journal *Nature*. In an accompanying article, Dr. Stephen Jay Gould, a paleontologist at Harvard University, wrote that the findings provide "no support for Cope's Rule as a preferential bias in the evolution of size."

Indeed, Dr. Gould and other scientists said the research revealed that evolutionary biologists had long allowed themselves to be deluded by a general bias in the way they often viewed nature. Like most people, they have been unduly impressed by bigness.

"Might not our conviction about the validity of Cope's Rule," Dr. Gould asked, "be a psychological artifact of singling out lineages that display size increase because we all know that 'bigger is better'?"

Dr. Gould in 1988 first spoke of the need for reconsidering Cope's Rule and in his latest book, *Full House*, published by Harmony Books, elaborated on weaknesses in some of the rule's supporting evidence. And in the *Nature* article, he wrote that the new findings also called attention to "another pervasive and lamentable bias of human reasoning: our tendency to focus on extremes that intrigue us, rather than full ranges of variation."

Such has been the case in studies of the horse. The horse is often cited as the classic example of Cope's Rule. But in fact, Dr. Jablonski pointed out, horses show a broad range of sizes through most of their evolutionary history—until the end, when all became extinct except for one of the larger lineages.

"The last survivor just happened to be a large one," he said. "If you connect the small starting point with the big final survivor, you seem to get a straight line of size increase, but the real pattern is much more complicated."

No doubt Cope was impressed by bigness because he was a vertebrate paleontologist working at the time that *Diplodocus, Stegosaurus* and other huge dinosaurs were being excavated in the American West. A wealthy Philadelphian, Cope financed his own fossil-hunting expeditions, usually in bitter competition with Othniel C. Marsh of Yale University, and was an impetuous genius who never shied from speculation.

When in 1987 he decided to test Cope's Rule, Dr. Jablonski concentrated on mollusks, the group of marine organisms he had already been studying. Rich collections of mollusk fossils were available, especially at the American Museum of Natural History in New York City, the Smithsonian's National Museum of Natural History in Washington and the United States Geological Survey in Reston, Virginia.

He further chose to examine late Cretaceous specimens because the mass extinction at the end of that period served as a sharp cutoff of normal evolutionary processes. At least half of the mollusk species became extinct then, though many of the clams and snails of that period have relatives living today. For his measurements, he examined adult sizes of each genus at the beginning and end of the study interval.

Analyzing the data, Dr. Jablonski found that 27 percent to 30 percent of the lineages showed a net increase in body size over the 16-million-year period. But just as many lineages displayed an overall body size decrease, while 28 percent showed an increase at both ends of the size scale. Only a few lineages showed a decrease in the total range of body sizes. A directional net increase in body size is thus no more frequent than an increase in the range of sizes among species, or even a net evolutionary size decrease.

In short, he said, the results are "very much in opposition to the classical Cope's Rule."

What are the implications for evolutionary biology? "We will have to be more skeptical about assertions that natural selection favors this or that body size," said Dr. McShea of Duke.

Under the influence of Cope's Rule, many scientists have assumed that large body size somehow bestowed critical evolutionary advantages: the attraction of mates, self-defense, resistance to temperature extremes and mobility over a greater foraging and hunting range. Consequently, larger organisms would have a better chance of passing on their genes, which would eventually translate into a pattern of ever-increasing body sizes within a lineage.

But in at least one fateful case, it may have paid to be small. Early mammals, many of them no bigger than shrews, managed to survive the Cretaceous mass extinction. Among the advantages often ascribed to small-bodied organisms is rapid reproduction and the ability to hunker down to avoid predation or survive environmental change. While big dinosaurs were dying, many small mammals hung on to evolve into scientists who would contemplate the merits of being small or big.

"My data show that although body size may be tremendously important in an ecological sense," Dr. Jablonski said, "there is no simple extrapolation to size being important in a long-term, large-scale evolutionary sense."

—JOHN NOBLE WILFORD, January 1997

The Wizard of Eyes: Evolution Creates Novelty by Varying the Same Old Tricks

ONE OF THE MOST FASCINATING yet awkward problems that have challenged biologists since Charles Darwin is how to explain the evolution of complex structures like the eye and the wing. How does a blind force like natural selection sculpture from almost nothing a supremely delicate mechanism like the inner structure of the ear?

The trick, so hard to understand, is one that apparently gives evolution no bother. The eye, for instance, is an organ that has evolved not once but perhaps dozens of times, since insects, octopuses, mammals and other groups all have eyes of quite different structure.

Scientists studying how the trick is accomplished at the level of genes and their protein molecules say they are finding a common thread in this varied tapestry: Evolution produces the strikingly new by tinkering with the old.

Animals do not evolve whole new suites of genes and proteins to build complex structures. Instead, a species appears to construct useful new features by adapting existing molecules to new uses. In some cases, researchers say, evolution simply changes the settings in the program by which an embryo develops into an adult, producing some powerfully novel results just by altering the timing and extent of use of existing genetic building blocks.

"We're now beginning to provide some answers to the questions Darwin raised," said Dr. Margaret McFall-Ngai, a developmental biologist at the University of Southern California at Los Angeles and one of the speakers discussing the evolution of novelty at a meeting on development and evolution at Indiana University in Bloomington. "We're finding that the organism is extremely resourceful. It has a set of cards to play and it can deal them up a bunch of different ways to come up with novel structures. It's changing our ideas about how these structures are formed."

Dr. Rudy Raff, director of the Institute for Molecular and Cellular Biology at Indiana University and an organizer of the symposium, said, "The most surprising and important discovery is the sameness. You see the same genes used again and again, always co-opted for other uses. If you just take what you have and reorganize it, you can make something new rather quickly, and I think that's the way evolution works. And that's one of the surprising things molecular biology has really revealed."

The puzzle of how an organ as exquisitely complex as an eye could first evolve was problematic enough that Darwin felt compelled to discuss it at length in his book *The Origin of Species* in a chapter entitled "Difficulties of the Theory."

"It's the problem of the incipient stages of useful structures," said Dr. Stephen Jay Gould, an evolutionary biologist at Harvard University who was the keynote speaker at the meeting. "We can all see what wings are for when they're complete wings. But just a few percent of a wing, what good is that? Darwin's general answer is that it must exist for other purposes, and that continues as the key theme in the evolution of novelty. You can't get complexity out of nowhere."

And indeed you don't, as researchers at the National Institutes of Health studying eye lenses are discovering.

The lens is a remarkable transparent structure filled with proteins called crystallins, which have been an object of study for more than a century. Found in great quantities in the lens and perfectly arranged to focus light, crystallins were long thought to be highly specialized lens proteins that evolved solely to allow animals to see.

But as researchers studying these proteins are discovering, the evolution of the lens may have been a lot simpler than was imagined and may begin to provide lessons in how such complex structures arise in general.

Many crystallins, scientists are finding, are actually the same proteins as the common housekeeping enzymes that govern the cell's basic chemical metabolism. One crystallin, for example, is also known from its humbler role as lactate dehydrogenase, an ordinary metabolic enzyme. Researchers say these proteins evolved first for their humdrum housekeeping tasks and only later were adapted to the entirely different and specialized task of building the lens.

"The story in the lens shows that you can evolve complex body parts by borrowing specialized proteins that are being used for one function for use in an entirely different function," said Dr. Joram Piatigorsky, chief of the laboratory of molecular and developmental biology at the National Eye Institute in Bethesda, Maryland, and a speaker at the meeting. "I'd rather expect that similar kinds of things are going on everywhere."

Dr. Charles Zuker, a molecular biologist at the Howard Hughes Medical Institute at the University of California at San Diego, said, "Why reinvent a molecule to fulfill a given function if you can simply recruit an existing player that's involved in another pathway? It wouldn't make sense to reinvent every molecule for every pathway that needs to be filled."

The double life of these proteins may, in fact, be crucial to their role in the lens. Many of the crystallins' original jobs outside the lens appear to have been to protect themselves and other proteins from damage under stress and over long periods of time. Researchers say this allows the crystallins to do double duty in the eye, where many proteins are produced before birth and never replaced. Not only can they function to bend light, but crystallins can also help maintain their own integrity as well as that of the other irreplaceable proteins that must function in the lens for a lifetime.

Dr. Graeme Wistow, a molecular biologist at the National Eye Institute, has studied the lenses of a wide variety of animals and reports further evidence of evolution's resourcefulness. "The lens is the interface with the outside world," he said. "So, for example, when our ancestors emerged from water onto land, it was necessary to change the lens's properties. In this and in other cases it seems that there have continued to be a lot of changes in the lens's protein composition."

Such changes in composition over time are reflected in the array of crystallin mixes found in the lenses of the many different animals so far studied, including fishes, humans, squids, shrews and hummingbirds. These mixes, Dr. Wistow suggests, are the result, at least in part, of the many different needs for vision in different circumstances—under water, on land, by day, by night. And as animals explore new visual worlds, more and different metabolic enzymes continue to be recruited into the eye for the ever-evolving needs of the lens.

Other skillful adaptations have been found by Dr. McFall-Ngai in the lens-like structures of light-producing organs in squids.

Preyed upon by fishes swimming below them in the water, the squids use bioluminescent organs, called photophores, to hide the shadows they cast while swimming in moonlit or starlit water. By casting a matching beam of light down below them, they effectively erase their telltale silhouettes.

While the eyes of these squids contain a lens that takes light from the environment and focuses it onto the retina, the photophore, built like an eye in reverse, contains a lens that diffuses the light produced within it out into the environment.

"We wanted to know, how do you make a transparent structure, a lens, somewhere else in the body, like in a photophore?" said Dr. McFall-Ngai. "How does the squid make this biochemically?"

It turns out that an eye-specific form of a housekeeping enzyme known as aldehyde dehydrogenase was one of the major crystallins filling the squid eye's lens. But unlike the squid eye, which makes its lens from the outer layers of skin tissue, the photophore makes its lens from muscle tissue. Consequently, biologists assumed it had evolved its lens independently and in an entirely different fashion.

Yet researchers found that the protein used to form the lens of the photophore was not the muscle-specific form of aldehyde dehydrogenase or any of the proteins typically found in muscles, but the peculiar form of aldehyde dehydrogenase used specifically in the eye lens. Instead of evolving a new lens protein to make a lens-like structure in the photophore, the squid simply reached into its existing bag of tricks.

"The lesson is that there's nothing new under the sun," said Dr. McFall-Ngai. "The genome is incredibly resourceful, and when posed with the question 'How do I achieve the development and evolution of this new structure?' there's more and more evidence that it just goes back to itself and says, 'Let's rearrange these genes in a different way in a different place to come up with what would appear to be something completely novel.'"

Researchers have even found that the genes that regulate the development of eyes have been widely used and reused across the animal kingdom. In a finding that researchers say surprised many, a Swiss group found that the same gene appears to control the development of both the multifaceted, kaleidoscopic eyes of fruit flies and the eyes of humans. Yet these two were considered classic examples of complex structures that had evolved independently.

Dr. Walter Gehring, a developmental biologist at the Biozentrum at the University of Basel, and his colleagues found that the fruit fly gene known as "eyeless," which regulates the development of the insect's eyes, shows striking similarity to the Pax-6 gene, which controls eye development in vertebrates, including humans. Once again, researchers studying the fine details of evolution's craftsmanship in the eyes of humans and flies have found not novelty but parsimony.

Researchers say the new findings promise to invigorate the study of development and evolution, a field whose potential they say has only just begun to be realized.

Dr. Raff, who is writing a book about new developments in the fast-growing field, said, "In the 1940's, genetics, paleontology and evolution were pulled together to synthesize a new way of understanding evolution. Now we can tie these different things together with development, especially with all the new molecular data. Evolution gets recreated every generation or two, and perhaps it's time again."

—CAROL KAESUK YOON, November 1994

6

THE
EMERGENCE
OF HUMANS

Of all the events in the fossil record, none is more fascinating than the emergence of humans. The still-unfolding story shows the limitations as well as the power of paleoanthropology, as this specialized branch of paleontology is called.

The raw material of paleoanthropology used to be extremely scarce, just a precious handful of skulls that could be laid out on a billiard table. But that was in part because those who hunted for early human fossils were digging in the wrong places. Assuming that the divergence between humans and apes must have occurred some 30 million years ago, they were searching for fossils in rock strata of that age.

But biologists looking at molecular clocks declared that the split between humans and chimpanzees, based on comparison of their DNA, took place much more recently, maybe just 5 million years ago. Now that paleoanthropologists have started to dig in rocks of that and younger ages, early human remains have begun to be more plentiful.

The last few years have seen a rich trove of discoveries that bear on the crucial dividing point between the human and the ape ancestries. Several more specimens have been discovered of *Australopithecus afarensis,* the hominid that lived from 3.6 million to 3 million years ago and is thought to be a direct human ancestor. Even more interesting, two apparently new species of even greater antiquity have been unearthed, *Australopithecus ramidus,* dated to 4.4 million years ago, and *Australopithecus anamensis,* which lived from 4.2 million to 3.9 million years ago.

Both species, like *afarensis,* walked upright, but lie much closer to the human-ape split. Further study of these specimens and the environment in which they lived may shed light on the origin of bipedality, or walking on two feet. Many paleoanthropologists feel that bipedality, since it freed the hands for tasks like toolmaking, was the critical behavior that shaped the evolution of human forebears.

One long-standing mystery in paleoanthropology, the kinship of Neanderthals and humans, seems at last to have been resolved. Neanderthals were a heavy-boned hominid species that lived in Europe and the Middle East from 300,000 to 30,000 years ago, co-existing for part of that time with modern humans, who emerged as recently as 100,000 years ago.

Some paleoanthropologists have argued that Neanderthals were a separate species that passed into extinction, probably at the hands of humans; others contend that humans are descended from Neanderthals. Surprisingly, the debate seems to have been definitively settled by the recovery of DNA from a Neanderthal fossil. The analysis shows that the Neanderthal lineage is four times older than that of man, Neanderthals being an earlier and separate offshoot from the hominid line and not a direct ancestor of humans.

As the following articles indicate, the pace of discovery in paleoanthropology is as brisk as ever and the field has seldom been more interesting.

———————————————

New Fossils Take Science
Close to Dawn of Humans

FOSSILS OF THE OLDEST human ancestors have been discovered in Ethiopia, where these ape-like creatures lived 4.4 million years ago on a forested floodplain. Not only do they represent an entirely new species, scientists said, but they also may well be the long-sought relatives that lived close to the fateful time when the lineages leading to modern apes and *Homo sapiens* went their separate ways.

"The discovery of these ancient fossils and their context signals a major step in our understanding of human origins," Dr. Tim D. White, a paleontologist at the University of California at Berkeley, said yesterday in an announcement of the findings.

Other scientists greeted the announcement with ringing endorsements. They said the evidence for identifying the new species was compelling. The species represented an enormous leap back of 800,000 years in reconstructing the pre-human fossil record. If genetic research by molecular biologists is correct, the find has taken scientists into that intriguing evolutionary time, estimated to be six million to four million years ago, when the apes and humans diverged.

The name assigned to the new species, *Australopithecus ramidus,* is a reflection of its presumed primal importance on the human family tree. In the Afar language of the region where the fossils were found, *ramid* is the word for root, and it applies to plants or people—thus humanity's root species.

Details about the fossils, excavated in the last two years at a site in the Ethiopian badlands called Aramis, are reported in the journal *Nature* in an article by Dr. White and his two principal colleagues, Dr. Gen Suwa, a paleontologist and expert in ancient teeth at the University of Tokyo, and Dr.

Berhane Asfaw, an Ethiopian paleontologist. They described the fossils as "the most apelike hominid ancestor known," something scientists have been seeking for two decades, a closer "link in the evolutionary chain of species between humans and their African ape ancestors."

In a commentary accompanying the report, Dr. Bernard Wood, a paleontologist at the University of Liverpool in England, wrote, "The metaphor of a missing link has often been misused, but it is a suitable epithet for the hominid from Aramis."

The fact that the fossils were found in sediments of a previously wooded environment could also be of profound significance in reinterpreting the early stages of human evolution. It may even be the most important immediate consequence of the research.

"The most exciting thing about this discovery is the ecological context," Dr. C. Owen Lovejoy, a paleontologist at Kent State University in Ohio, said in an interview.

The assumption had been that when the climate changed and forests gave way to grasslands, the forces of natural selection favored those apes that could walk upright and adapt to the open country. These adaptations led to the first hominids. If the new species did indeed evolve in a forest, scientists will have to rethink these assumptions and, as Dr. White said, consider that it "was not the savanna that forced us along the evolutionary road."

As yet, there is only indirect evidence that these creatures, which were about the size of chimpanzees, were able to walk upright. Most of the fossils are of teeth, jaws, a cranial base and some arm bones. One of the objectives of the next excavations, soon to begin, is to search for pelvic, knee and foot bones that should be more revealing of the species' walking abilities.

By shaking up conventional wisdom and extending the chronology of human ancestry, Dr. Lovejoy said, "This is the most exciting thing to happen since Lucy."

Until the new discoveries, the earliest known direct human ancestor was *Australopithecus afarensis,* the first and most famous specimen of which was found in 1974 and nicknamed Lucy. This partial skeleton was dated at 3.2 million years, and other finds showed that the same species lived between 3.6 million and 3 million years ago, and perhaps even earlier, though the evidence is sparse and not well established. Mary Leakey, the noted Kenyan paleontologist, uncovered the fossil footprints in 1977.

This trail of footsteps embedded in hardened volcanic ash at the Laetoli site in Tanzania was made by two or three individuals walking with an upright, human-like gait. This discovery represents the earliest indisputable evidence of hominid bipedality, and it is usually attributed to *Australopithecus afarensis,* the species to which the Lucy skeleton belonged.

Although there are almost as many hypothesized human family trees as there are paleontologists, scholars generally agree that *A. afarensis* is the common ancestor to two subsequent lines of evolution. One is the heavy-jawed, small-brained australopithecines, which became extinct about one million years ago. The other line began with the emergence of the genus *Homo* about 2.5 million years ago, about the same time the first stone tools were made and used. The most immediate human ancestor was *Homo erectus,* from 1.8 million years to perhaps a few hundred thousand years ago.

After two decades of research, scientists recognized that the Lucy species still left them with a substantial gap in the initial evolution after the ape-hominid divergence. The search for a species ancestral to *A. afarensis* has been a central goal of fossil hunters.

Fossils of 17 individuals of the new species, *A. ramidus,* were excavated about 140 miles northeast of Addis Ababa and 45 miles south of Hadar, where the Lucy skeleton was found. Dr. Suwa made the initial discovery on December 17, 1992. As he walked across the barren ground, his eye was caught by the glint of a molar tooth among the desert pebbles.

"I knew immediately that it was a hominid," Dr. Suwa said. "And because we had found other ancient animals that morning, I knew it was one of the oldest hominid teeth ever found."

Analyzing the fossils, the scientists realized that these individuals were more like chimpanzees, the apes that are the closest living relatives of humans, than were members of the afarensis species. But the reduced size and different shapes of certain teeth, particularly the canines and a lower first deciduous molar, one of the so-called milk teeth, showed that these creatures were more primitive than afarensis but had evolved from the same apes that had been their ancestors.

These were the remains of a species, Dr. Wood said, that "lies so close to the divergence between the lineages leading to the African apes and modern humans that its attribution to the human line is metaphorically, and literally, by the skin of its teeth."

Dr. Asfaw said: "The short cranial base and the hominid shapes of the canine and elbow show us that this species had already split from the apes. It had started to evolve toward human beings."

Since scientists had been expecting to find a species ancestral to afarensis, but much more primitive, said Dr. Donald C. Johanson, president of the Institute of Human Origins in Berkeley and, along with Dr. White, a discoverer of the Lucy skeleton, "There are no real surprises in the anatomy."

Dr. Johanson said the findings provided strong fossil support for the genetic studies suggesting that the split between apes and humans occurred relatively recently, perhaps no more than six million years ago. Many paleontologists had long held out for a much more ancient divergence, perhaps as much as 20 milion to 15 million years ago. Scientists said there could still be other transitional species closer to the common ancestor of apes and humans.

In another report in *Nature*, a team of geologists and anthropologists, led by Dr. Giday Wolde Gabriel of the Los Alamos National Laboratory in New Mexico, described how the fossils lying under volcanic ash and glass were dated and noted that an abundance of fossil wood and arboreal seeds was identified. This and a substantial number of monkey fossils were offered as evidence that the new species had "lived and died in a woodland setting."

Like several scientists who have been challenging traditional thinking, Dr. Andrew Hill, a Yale University paleontologist, said: "This supports the hunches we have had that savannas were not the major element in the origins of hominids. I would be happier if we knew it was a biped."

For other scientists to accept this revisionist view, the explorers at Aramis must come up with stronger evidence that *A. ramidus* either could walk upright or at least appears to have been evolving such a talent. Dr. White said the smaller canine teeth of the new species were consistent with characteristics of other hominids that had altered their social structure and were beginning to carry infants and food with their forearms, the first stage of upright walking. Other indirect evidence included characteristics of the cranial base that were different in some respects from apes, indicating that it had evolved in ways associated with upright walking.

Somewhat more direct evidence, Dr. White said, were the similarities in the arm bones of *A. ramidus* and the Lucy species. "This is not the arm of a knuckle-walker," he said, "but I want to hold the question of locomotion until more fossils are found."

—JOHN NOBLE WILFORD, September 1994

Tiny Foot Bones May Show
a Giant Leap for Mankind

A VOLCANO in what is now Tanzania erupted, scattering fine-grained ash over the Serengeti Plain. Rain followed. Then across the plain covered with damp ash walked two individuals, possibly accompanied by a third, smaller figure, headed due north. They left a trail of footprints that hardened and remained preserved under a covering of more ash and silt until their sensational discovery in 1977.

These footprints preserved in volcanic ash show that human ancestors in Africa were walking upright as early as 3.7 million years ago, making this behavior one of their first and most distinctive departures from all other primates. But was walking the only way these protohumans got about? Were they still climbing trees, in the manner of chimpanzees, for safety, sleeping and gathering fruits and nuts?

These questions have long puzzled and divided anthropologists studying crucial evolutionary changes in the early hominids after their divergence from the apes. Now the discovery of four foot bones in South Africa has produced the best fossil evidence yet for how the ancestors of modern man moved about, and may yield compelling answers or simply rekindle the flames of old debates.

The four little bones, rather than settling the debate, have already caused more fur to fly among the small group of contending experts who study human origins.

In a discovery paleontologists describe as the most important in southern African hominid exploration in recent decades, the four bones formed the instep and the beginning of the great toe of a hominid that may have lived as much as 3.5 million years ago, presumably a member of the *Australopithecus* genus. This is the oldest set of connected foot bones of any hominid and the earliest evidence, by more than half a million years, for the existence of such ancestors south of tropical Africa.

But of greatest importance to the history of human evolution, analysis of the bones, dubbed Little Foot, has revealed that this hominid combined human-like and ape-like foot characteristics. The weight-bearing heel and the springy arch of the foot were unquestionably adapted for upright walking much like modern humans, said paleontologists who examined the fossils, while the great toe was set at a wide angle to the other toes and was highly flexible, presumably capable of grasping and climbing. In particular, the shape of the joint forming the ball of the foot indicated that the big toe could rotate inward like the opposable human thumb, useful in tree climbing.

In a report of their research published in the journal *Science,* Dr. Ronald J. Clarke and Dr. Phillip V. Tobias, paleontologists at the University of Witwatersrand in Johannesburg, said the foot fossils were "the best available evidence that the earliest South African australopithecine, while bipedal, was equipped to include arboreal or climbing activities in its locomotor repertoire. The exact proportion of its activities spent on the ground and in the trees is at present indeterminate."

The bones were discovered in 1980 but their significance was not recognized until last year. They were found in the deepest part of sediments in the Sterkfontein Cave, which is near Johannesburg and has been the site of numerous early human discoveries. The bones probably belonged to an early member of *Australopithecus africanus* or another early hominid species, the paleontologists said. They are the first connected bones found from the same foot of a single individual of such creatures.

"This may be the kind of foot that made Mary Leakey's footprints," Dr. Tobias said in a telephone interview from Johannesburg.

Most paleontologists had seen in the afarensis fossils clear evidence for upright walking, but several anatomists, notably Dr. Jack T. Stern and Dr. Randall L. Susman of the State University of New York at Stony Brook, have argued that the foot and ankle remains "reveal to us an animal that engaged in climbing as well as bipedality."

Dr. Bernard Wood, a paleontologist at the University of Liverpool in England, has pointed out that Lucy's limb proportions and skeleton suggest she was neither predominantly tree-living nor fully upright. Writing in the *Cambridge Encyclopedia of Human Evolution,* he said that Lucy and her relatives might have "spread out to forage on the ground in the day, and then congregated, perhaps in caves or trees, at night."

Dr. Clarke and Dr. Tobias acknowledged that their findings supported the conclusions of Dr. Stern and Dr. Susman that hominids of this period represented an "intermediate degree of adaptation" in locomotion.

Predictably, Dr. Susman in an interview praised the report as a "conceptually and theoretically very compelling paper." If the South African fossils are indeed of the africanus species, not afarensis, the findings show, he said, that it is probably "the blueprint for all hominids" that transitional species were part arboreal quadrupeds and part ground bipeds. "Fully committed bipeds," he said, probably did not emerge until *Homo erectus* about 1.5 million years ago.

A leading proponent of completely bipedal early hominids, Dr. C. Owen Lovejoy of Kent State University in Ohio, professed to be unshaken by the findings. To him the Laetoli footprints had shown that the great toe of hominids had already moved in line with the other toes and was no longer opposable, indicating an individual that had clearly walked away from any four-legged, tree-living past.

In an accompanying article in *Science,* Dr. Lovejoy was quoted saying the conclusions of Dr. Clarke and Dr. Tobias were "patently absurd." He also said that the australopithecine hip, knee and spine had by this time been adapted for an upright life and so to ignore that evidence in favor of one foot joint was, in his words, "mechanically and developmentally naive."

When these comments were read to him, Dr. Susman said: "That's just Owen blustering. He still thinks hominids have to be a biped and nothing else."

Dr. William Kimbel, a paleontologist at the Institute of Human Origins in Berkeley, California, said that it was "conceivable to have primitive features like the divergent big toe without implying that they are maintained for the function of arboreal activity."

In other words, he said, the new findings were not likely to satisfy all scientists because of their differing views on interpreting the relationships between morphology and function, particularly the function of retained ape-like characteristics.

Other paleontologists generally agreed that the research was important but expressed reservations about the dating of the fossils at 3 million and possibly 3.5 million years old.

The age of Little Foot could provoke other controversies. If it is 3.5 million years old and from the africanus species, Dr. Susman noted, this would mean that it could not be a descendant of afarensis, which many paleontologists have considered the common ancestor of all subsequent australopithecine and eventually *Homo* species.

"There's enough in this paper to get everybody's blood boiling," Dr. Susman said.

—JOHN NOBLE WILFORD, July 1995

New Fossils Reveal the First of Man's Walking Ancestors

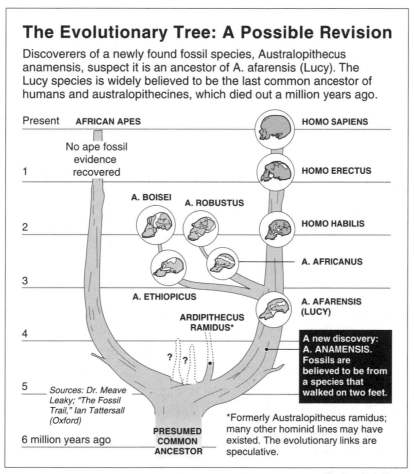

The Evolutionary Tree: A Possible Revision

Discoverers of a newly found fossil species, Australopithecus anamensis, suspect it is an ancestor of A. afarensis (Lucy). The Lucy species is widely believed to be the last common ancestor of humans and australopithecines, which died out a million years ago.

Present AFRICAN APES HOMO SAPIENS

No ape fossil evidence recovered

1 HOMO ERECTUS

A. BOISEI A. ROBUSTUS

2 HOMO HABILIS

A. AFRICANUS

3

A. ETHIOPICUS A. AFARENSIS (LUCY)

ARDIPITHECUS RAMIDUS*

4 A new discovery: A. ANAMENSIS. Fossils are believed to be from a species that walked on two feet.

? ?

5 Sources: Dr. Meave Leaky; "The Fossil Trail," Ian Tattersall (Oxford)

6 million years ago PRESUMED COMMON ANCESTOR

*Formerly Australopithecus ramidus; many other hominid lines may have existed. The evolutionary links are speculative.

The New York Times

FOSSIL DISCOVERIES in Kenya have revealed a new species of human ancestors that lived four million years ago, were hardly any bigger than chimpanzees, bore striking resemblances to both apes and evolving man and, perhaps most significantly, were already standing erect and walking on two legs much like modern humans.

Leg and arm bones found near Lake Turkana, paleontologists said, provide the earliest direct and unambiguous evidence for upright walking, or bipedalism, by any members of the human family tree. Until now, the oldest evidence has been the 3.6-million-year-old footprints at Laetoli in Tanzania, together with tiny foot bones from South Africa that are about the same age.

By placing the emergence of bipedalism further back by about half a million years, the discovery supported the growing impression that this novel mode of locomotion might be the defining adaptation that first set human ancestors and their close relatives, known collectively as hominids, apart from the quadrupedal apes. From genetic studies comparing modern apes and modern humans, molecular biologists have established that the split between the two lineages occurred seven million to five million years ago.

The discovery also crowds and probably complicates the base of the hominid family tree. It could provoke a new round of controversy among scientists trying to reconstruct human genealogy.

For two decades, the earliest known hominid was a species represented most famously by the Lucy skeleton. *Australopithecus afarensis* appeared to be the sole hominid species from 3.9 million to 2.9 million years ago. Reading the DNA clock in the genes based on guesses about the rate of mutation, molecular biologists started asserting that the split from the apes was far more recent than most paleontologists believed—a mere 5 to 7 million years ago. Thus inspired, paleontologists searched more diligently for signs of earlier hominid life by looking in Africa for fossil-bearing sediments of this critical age.

With the discovery near Turkana, two new species became known in 1994. A 4.4-million-year-old species from Ethiopia was identified and described; first lumped in the *Australopithecus* genus, it was recently reassigned to a new genus and given the name *Ardipithecus ramidus*. These hominids may have walked upright, but no direct evidence for this has yet been reported.

Now, in a report published in the journal *Nature*, a team led by Dr. Meave Leakey of the National Museums of Kenya announced the discovery of the second new hominid, which has been named *Australopithecus anamensis*. The fossils of 21 specimens, including teeth and parts of the jaw, were found at two sites near Lake Turkana, Kanapoi and Allia Bay. The name *anam* means lake in the Turkana language.

If the name of the principal discoverer has a familiar ring, it may be because she is the wife of Richard Leakey, the prominent Kenyan fossil hunter who has left the field because of a crippling airplane accident and new interests in wildlife preservation and politics. He is the son of Mary Leakey and the late Louis S. B. Leakey, whose many discoveries this century established East Africa as the ancestral grounds for many human forerunners. The British-born Meave Epps Leakey, following in the family tradition, is an experienced paleontologist with a doctorate in zoology and years of fossil hunting behind her.

The co-authors of the *Nature* report are Dr. Alan Walker, an anatomist at Pennsylvania State University who has worked for years with the Leakeys, and two geologists, Dr. Craig S. Feibel of Rutgers University in New Brunswick, New Jersey, and Dr. Ian McDougall of the Australian National University in Canberra. Their research was supported by the National Science Foundation and the National Geographic Society.

Geologists have dated the new specimens at 4.2 million to 3.9 million years old, making them intermediate in time between the ramidus species discovered by Dr. Tim White of the University of California at Berkeley and the later afarensis species, whose Lucy fossils were found in 1974 by Dr. Donald C. Johanson, a paleontologist now at the Institute of Human Origins in Berkeley.

The new species seemed to be a blend of primitive and advanced characteristics. The jaw, with its shallow palate, large canines and placement of teeth, harks back to the apes. So do the extremely small ear openings in the skull. But the tooth enamel is much thicker than that of the apes or the ramidus hominid, thus resembling a feature of later hominids in the *Homo* lineage. And the tibia, or shinbone, and humerus, the upper-arm bone, are especially advanced. In one of the most telling clues to bipedality, the upper end of the shinbone is shaped to bear more weight than a four-legged animal would require.

Dr. Ian Tattersall, a paleontologist at the American Museum of Natural History in New York City, said the new fossils seemed to be "more plausibly hominid than ramidus is" and should advance understanding of when and where upright posture first evolved.

Dr. Johanson, often a rival of the Leakeys in finding and interpreting early human fossils, endorsed the conclusions drawn from the new finds.

Reached by telephone in Berkeley, Dr. Johanson said the anamensis species announced today "does seem to have a series of features different from afarensis," the species represented by the Lucy skeleton, and so is worthy of a distinctive place on the family tree. The species "fits the evolutionary trajectory of these early australopithecines," he said, and it emphasizes how the 4.4-million-year-old ramidus "seems to stick out as not being part of" this trajectory and may, instead, be "an ancestor to later apes rather than later hominids."

After examining the mixture of primitive and advanced characteristics of the anamensis species, Dr. Leakey and her colleagues wrote that the evidence "shows this species to be a possible ancestor to *Australopithecus afarensis*." But they, too, questioned whether the earlier ramidus hominid was a direct ancestor of either anamensis or afarensis.

Because the ramidus fossils appeared to be so much more primitive than anamensis, Dr. Leakey's team decided it was more likely that this earlier hominid was a sister species to all later hominids and thus represented a different branch of the hominid tree.

"Either evolution occurred very quickly for ramidus to be a direct ancestor of anamensis, or else Tim's fossils are on a side branch," Dr. Walker said in a telephone interview, referring to Tim White, the ramidus discoverer.

In a letter to *Nature* in May, Dr. White acknowledged that the ramidus hominid is likely to be a sister branch in explaining why he was assigning it to an entirely new genus. But he has also been quoted as still believing that ramidus was an ancestor to the hominids leading to afarensis and eventually the *Homo* lineage.

Dr. White has found more ramidus fossils but has not yet published his results. "We're on tenterhooks waiting to see what Tim has," Dr. Walker said.

In any event, paleontologists generally agreed that the new fossils represented a distinct species and that this added to the growing body of evidence for a great diversity among hominids around four million years ago. They could picture an early period when several hominid lines co-existed, and only one led eventually to humans, with the others dying out long ago.

"*A. anamensis* is seen as a possible ancestor to *A. afarensis*, although it is recognized that four million years ago there may have been a number of newly emergent hominid species with variations based on the novel bipedal adaptation," Dr. Leakey and her colleagues wrote, noting that there "is no theoretical reason why there should be only one early hominid species at any one time." In their view, ramidus is more likely to have been one of the ill-fated lines and anamensis an occupant of the successful line.

—JOHN NOBLE WILFORD, August 1995

The Transforming Leap, from Four Legs to Two

THOSE WERE VENTURESOME STEPS for some ape-like creatures long ago in Africa. Dropping out of trees, they essayed a novel means of locomotion, for reasons that elude paleoanthropologists. These primates may have sought to reach higher in foraging, see farther over tall grasses, reduce the exposure of their bodies to the searing tropical sun or extend their range beyond the forest to the open savanna. Perhaps they needed to free the hands for carrying food or infants over long distances.

In any case, driven by need or the lure of new opportunity, in their struggle for survival these creatures found some saving advantage in a new way of walking. Instead of scampering about on all fours, as usual, they stood upright and, gradually and no doubt unsteadily at first, began to walk on their hind limbs. Small bipedal steps for apes, and as it would turn out, a giant leap for mankind.

Anthropologists and evolutionary biologists are now agreed that upright posture and two-legged walking—bipedality—was the crucial and probably first major adaptation associated with the divergence of the human lineage from a common ancestor with the African apes. Once they had thought the development of a large brain or the making and use of stone tools was the pivotal early evolutionary innovation setting human ancestors, the hominids, apart from the apes. But these came much later, long after the transforming influences of bipedality.

Upright walking required profound changes in anatomy, particularly in the limbs and pelvis, and these were passed on to modern humans. It eventually put limits on the size of infants at birth and thus created the need for longer postnatal nurturing, with sweeping cultural consequences. It may have had a bearing on human sexuality and the development of family life.

And it certainly opened the way for later toolmaking, some 2.5 million years ago, and probably set the stage for the eventual enlargement of the hominid brain, not before two million years ago.

"The fundamental distinction between us and our closest relatives is not our language, not our culture, not our technology," said Richard E. Leakey, a paleontologist and son of Mary Leakey and the late Louis S. B. Leakey, renowned fossil hunters in Kenya. "It is that we stand upright, with our lower limbs for support and locomotion and our upper limbs free from those functions."

If there was any lingering question about the early manifestation of bipedality in human evolutionary history, it has been erased by recent fossil discoveries: a 4.4-million-year-old hominid from Ethiopia, found last year, and a 4.2-million-year-old hominid near Lake Turkana in Kenya, recently reported. The newly identified species were older, more primitive and much more ape-like than any hominids known before, but already they were bipedal—certainly the younger one, probably the other as well.

"This gets close to the hypothesized time of splitting of the ape and human lineages," said Dr. Alan Walker, an anatomist at Pennsylvania State University who specializes in early human studies. He and Dr. Meave Leakey, a paleontologist at the National Museums of Kenya, who is Richard Leakey's wife, discovered the Turkana fossils, to which they have given the new species name *Australopithecus anamensis*.

Dr. Tim D. White, a paleontologist at the University of California at Berkeley, excavated the 4.4-million-year-old hominid, so different from anything seen before that it has been assigned to an entirely new genus as well as species—*Ardipithecus ramidus*. Dr. White has yet to assemble and analyze the pelvis and lower limb bones, but he has inferred from other evidence that these creatures probably had an erect posture for walking. They resembled apes more than even anamensis does, showing primitive qualities that might be expected of creatures more than five million years old.

"With ramidus," he said, "we are very, very close to the hominid-ape split, surprisingly close."

Molecular biologists, comparing the DNA of modern humans, chimpanzees and gorillas, estimate that the decisive separation of hominids from the ape line occurred seven million to five million years ago. Until now, the earliest and most evocative evidence for hominid bipedality was the footprints embedded in 3.7-million-year-old volcanic ash at Laetoli in Tanzania. They were presumably made by members of the species *Australopithecus afarensis,* until the recent discoveries the earliest known hominid group. The most famous afarensis skeleton, of a small female nicknamed Lucy found in 1974 at Hadar in Ethiopia, bore indisputable limb and pelvic evidence of bipedality.

The new findings thus pushed back the time for the emergence of bipedalism at least half a million years earlier than previously known, perhaps several hundred thousand years more than that. They have encouraged scientists in the belief that they will someday fill the fossil gap all the way back to the hominid-ape divide. They have already afforded a glimpse of hominids that were so ape-like in nearly every respect as to accentuate their one human-like trait: upright walking. More than ever, paleontologists say, it is clear that the Rubicon early hominids crossed was not the large brain or toolmaking, but bipedality.

Now the challenge—one of the ultimate questions in the study of human origins—is to understand why the earliest hominids stood up. "Bipedalism is a fundamental human characteristic," said Dr. Bernard Wood, a paleontologist at the University of Liverpool in England, "yet virtually nothing is known about its origins."

Almost all other mammals stand and walk or run on four limbs. Those that stand on two feet have quite different postures and gaits from humans. Kangaroos hop on their two feet. Some monkeys, natural quadrupeds, may occasionally walk bipedally, especially when carrying food. Chimpanzees, the closest relatives to human beings, are capable of brief bipedal walks, but their usual means of getting about on the ground is knuckle-walking—standing on their hind legs but stooping forward, resting their hands on the knuckles rather than on the palms or fingers, which are adapted for climbing and grasping in the trees. Of the primates, only humans are fully developed bipeds.

As scientists have learned to reconstruct ancient climate from cores drilled in the ice of Greenland and sediments on the sea floor, one of the favorite explanations for the transition to bipedality has centered on drastic environmental changes that swept Africa more than five million years ago. By that time, Dr. Elisabeth S. Vrba, a paleontologist at Yale University, has pointed out, global climate had become significantly cooler and drier. As it did, grasslands in sub-Sahara Africa expanded and rain forests contracted, shrinking the habitat where tree-dwelling primates lived and foraged.

Another factor possibly upsetting the East African environment at the time was the region's unsteady terrain. Dr. Yves Coppens, a paleontologist at the College of France in Paris, contends that a seismic shift, recognized by geophysicists, deepened the Rift Valley, which cuts through Ethiopia, Kenya and Tanzania. The sinking of the valley produced an upthrust of mountains, leaving the land west of the valley more humid and arboreal, while the east became more arid and dominated by savanna.

As a result, he hypothesizes, the common ancestors of the hominids and the chimpanzees found themselves divided. Those adapting to the humid west evolved into the chimpanzee family. Those left in the east, Dr. Coppens wrote in 1994 in *Scientific American,* "invented a completely new repertoire in order to adapt to their new life in an open environment."

In any event, according to the hypothesis, at least one type of these primates responded to the environmental crisis by venturing more and more into the open grasslands, looking for food, but retreating to nearby trees to escape predators and sleep at night. To move about more efficiently, perhaps also to keep a lookout above the grasses for distant food or predators, these primates began standing up and walking on two legs. Their success presumably improved their chances of surviving and passing on genes favoring this unusual stance and gait, leading eventually to bipedal hominids.

Critics have pecked away at the hypothesis. Some contended that the mechanics of two-legged locomotion were not energy efficient, compared with those of four-legged creatures like dogs, horses and the big cats. After studying the question, Dr. Henry McHenry and Dr. Peter Rodman, paleontologists at the University of California at Davis, established that bipedality was indeed an effective way of covering a large amount of territory in the foraging for dispersed plant goods, especially for hominids. Moreover, hominids were close relatives not of horses or lions, but of primates whose particular form of quadrupedality was adapted mainly to arboreal locomotion and was already modified for a semi-erect posture under some circumstances.

Walking on two legs, in the succinct conclusion of the two scientists, was "an ape's way of living where an ape could not live."

One problem with the environment hypothesis has arisen with the new fossil findings. Both the Leakey-Walker anamensis and the White ramidus bones were found in areas that were once densely wooded, not savanna. And by analyzing carbon residues in ancient soils from the Rift Valley, Dr. John D. Kingston, a Yale anthropologist, recently determined that for the last 15 million years the mix of forest and grasslands was much the same as it is today, which also raises questions about an ecological crisis underlying the change in hominid locomotion.

Proponents of the hypothesis are not backing down. The soil measurements, they contend, may not be precise enough to reflect significant but short-term changes in vegetation. Nor do they think that the wooded setting in which the two sets of fossils were found necessarily represents the environment in which the species both lived and foraged.

In fact, other evidence indicates that for a million years or more early hominids probably had the best of two worlds, combining efficient bipedal locomotion on the open ground with the grasping and climbing abilities of arboreal quadrupeds. They may have foraged on two legs and rested or hid up in the trees. In July, South African paleontologists reported finding "a locomotor missing link" in the hominid fossil record. The foot bones of a species that lived as much as 3.5 million years ago, they said, included a weight-bearing heel used for two-legged walking with a big toe capable of grasping, much like a chimpanzee's.

If the origin of bipedality was related to prospecting the opening grasslands, Dr. Peter E. Wheeler, a physiologist at Liverpool John Moores University in England, suggests that there may have been another contributing factor. The early hominids, he notes, might have found the heat there especially stressful. Most animals living on savanna can let their body temperature rise during the day without wasting scarce water by sweating. They have built-in ways of protecting the brain from overheating. Not so humans, and presumably their distant ancestors. The only way they can protect the brain is by keeping the whole body cool.

Then perhaps the hominids stood up to keep cool. From his studies with a scale model of the Lucy skeleton, Dr. Wheeler found that a quadrupedal posture would have exposed the body to about 60 percent more solar radiation than a bipedal one. Standing tall thus might result in a substantial reduction in water loss. And the upright body could also catch the cooler breeze above the ground.

The new posture, Dr. Wheeler also said, might explain the evolution of the human as the naked ape. It obviated the need for complete body hair. Hair not only keeps heat in, it keeps it out. By standing upright, hominids had less need of hair to keep heat out, except on the head and shoulders. The advantages of height might also explain the evolution over time of taller hominids.

Dr. Ian Tattersall, an evolutionary biologist at the American Museum of Natural History in New York City, finds the cooling hypothesis "particularly attractive," if not the whole story. Dr. Wood of the University of Liverpool wonders if it really explains the origin of bipedalism, or merely explains why bipedalism was advantageous when hominids routinely foraged in more open habitats.

Dr. Kevin D. Hunt, an anthropologist at Indiana University in Bloomington, has offered another possible explanation. Observing chimpanzees in the field, he saw that their bipedalism was related to feeding. On the ground they stood on their hind legs to reach for fruit in trees. They also stood up on branches to grab food on a higher branch. This suggested that hominids might have adapted to a form of bipedalism long before they came down out of the trees.

Many such ideas are too narrow to account for something as broadly transforming as bipedality, in the opinion of Dr. C. Owen Lovejoy, an anatomist at Kent State University in Ohio who is a specialist in research on the origins of human locomotion. Instead, he has proposed a behavioral explanation with such sweeping implications that if correct, it would amount to a kind of grand unified theory of early hominid evolution.

Any trait that proves to be so advantageous that it is passed on to succeeding generations through natural selection, Dr. Lovejoy points out, almost always has some direct bearing on the rate of reproduction. Sex, that is, may have been the motivating force.

Upright walking, according to his hypothesis, began in the relative safety of the forest floor, not on the more perilous open terrain. Female hominids, restricted by the demands of infant care, spent most of their time gathering fruit and insects over a limited range. Their rather poor nutrition probably accounted for the slow maturation of the young and a low reproduction rate of one child only every four or five years. And as long as the female was nursing, she was unavailable for copulation.

So there could have been an incentive for males to free their hands for carrying food, especially nuts and animal protein found on their wider-ranging foraging. They could have brought the food back and exchanged it for sex, as anthropologists have observed pygmy chimpanzees doing in Africa today. This could be the basis for some kind of long-term bonding, perhaps the forerunner of modern human mating practices. In this way, the female would spend less time in search of food and more time and energy caring for her young. The children's chances of survival would improve, and the females would presumably resume ovulating somewhat earlier, and thus be sexually receptive. And the male that had been bringing home the bacon would be the favored mate.

Accordingly, Dr. Lovejoy contends, "The males of such pairs were most successful if competently bipedal and capable of proficient provisioning." They would be more likely to pass on their genes to later generations, thus establishing bipedality as the hominid locomotion and improving hominid prospects by significantly improving reproduction rates.

Dr. Lovejoy goes further to say that such behavior related to two-legged walking may have contributed in the long run to other peculiar human traits. Other primate females display pronounced genital swellings when they are ovulating. The loss of external signs of ovulation in humans, he said, could have been related to the advantages of fidelity. Likewise, human females are the only primates with permanently enlarged breasts, which in other animals would be an announcement that the female was not ovulating.

"Only the male that consistently copulated with a female [with hidden ovulation] would have a high probability of fathering her offspring," he said.

Of the Lovejoy hypothesis, Dr. Tattersall said: "It makes a nice story. But I don't think many people are convinced. We know so little about the lifestyles of those early hominids."

In his book, *The Fossil Trail: How We Know What We Think We Know About Human Evolution*, published by Oxford University Press, Dr. Tattersall favors the explanation that upright bipedalism "was intimately tied up with a change in climate and environment, that it somehow represented a response to the shrinking and fragmentation of formerly extensive forests in the part of Africa lying in the rainshadow to the east and south of the great domed-up Rift Valley."

These open spaces were a source of roots, shrubs, grasses and even the carcasses of animals killed by predators, he wrote, "a whole range of food resources for creatures with the wit to exploit them."

But in an interview, he conceded that scientists may never know for sure what made hominids stand up and walk on two feet. "We will always be driven to speculate," he said, "and, hopefully, our speculations will become more informed and insightful."

—JOHN NOBLE WILFORD, September 1995

Fossil Family: Big Guys
Join Delicate Lucy

THE DISCOVERY of more fossils related to the famous skeleton nicknamed Lucy has provided strong evidence that these earliest known hominids varied strikingly in body size between males and females and lived over a wide geographic range but were all members of a single species.

Scientists say the 3.4-million-year-old fossils from Ethiopia could settle a lingering debate over whether many of the specimens identified as *Australopithecus afarensis* were correctly categorized as a single species or perhaps represented two species of different sizes, ranges and behavior.

Hominids are primates that diverged from the ape lineage and include humans and their immediate ancestors. The *A. afarensis* species to which Lucy belonged is generally thought to be the common ancestor of other australopithecines, which became extinct a million years ago, and of the *Homo* line leading to modern humans.

In the journal *Nature,* scientists led by Dr. Tim D. White, a paleontologist at the University of California at Berkeley, reported that the new specimens were found in 1990 in the Maka area of the Middle Awash River Valley, about 50 miles south of the site of Lucy's discovery in 1974. In 1979, Dr. White and Dr. Donald C. Johanson, now director of the Institute of Human Origins in Berkeley, described Lucy and the related fossils as a new species most likely ancestral to all later hominids.

The Maka fossils are the first major addition to the hominid record for this early period since the 1970's.

The scientists said analysis of several leg and arm bones and jawbones showed that the Maka individuals were remarkably similar anatomically to the Lucy specimens from the Hadar site in Ethiopia, and also to generally larger individuals found farther to the south, at Laetoli, in Tanzania.

The marked differences in sizes between the diminutive Lucy, who stood no more than three and a half feet, and other specimens as much as a foot taller prompted much of the criticism about lumping the specimens in one species.

"This new mandible pretty much wipes out this argument," Dr. White said in a telephone interview. "We are seeing exactly what you would predict for a single species with a wide range of body sizes."

Dr. Eric Delson, a paleontologist at the American Museum of Natural History in New York and Lehman College, said the new findings supported the prevailing view. "Nearly everyone now accepts that there was a single species, but a highly dimorphic one," he said.

A large difference in body size between males and females is called sexual dimorphism. The new research, scientists said, confirmed earlier indications that australopithecine males could be almost twice as heavy as females. Such proportional differences are much more similar to the dimorphism of gorillas than to modern humans. On average, human males are about 15 to 20 percent heavier and 5 to 12 percent taller than females.

The Maka fossils, the scientists said, showed significant variations in tooth and body size among specimens found in a small area, which indicated that the size difference was not a matter of geography or species differences.

Analysis of the fossils also confirmed previous research that these early hominids walked on two legs and showed no signs of walking on four limbs or climbing trees. This reinforced interpretations that the 3.7-million-year-old fossil footprints at Laetoli were made by a bipedal afarensis and not by some other hominid species.

The scientists said the Maka fossils provided "powerful support for the interpretation of A. afarensis as a single, ecologically diverse, sexually dimorphic, bipedal primate species whose known range encompassed Ethiopia and Tanzania."

Dr. White said: "This makes Maka the third major site yielding afarensis. This makes the argument weaker and weaker that there is a hominid out there we haven't found yet sharing time and geography with afarensis."

The richness of the Maka site was recognized more than a decade ago, but exploration in Ethiopia had been sporadic because of political upheavals and periodic moratoriums on excavations by foreign scientists. Scientists working with Dr. J. Desmond Clark of the University of California at Berkeley are still there looking for more relatives of Lucy.

—JOHN NOBLE WILFORD, November 1993

Skull in Ethiopia Is Linked to Earliest Man

THE FIRST reasonably complete skull of the species made famous by the discovery of the headless Lucy skeleton has been found near the bank of a dry riverbed in Ethiopia's arid badlands.

The skull, with its ape-like heavy brow, jutting jaw and small braincase, is apparently that of a large male that lived three million years ago.

The remarkable find, which fills a serious gap in understanding early human evolution, gives a face to the species. Without a skull, scientists had not been sure what these creatures looked like or exactly what Lucy's position was in the human lineage.

The discovery could thus settle some of the hotly debated issues over whether the varied fossils from this time, between 3.9 million and 3 million years ago, actually belonged to a single species, known as *Australopithecus afarensis* and considered the common root of the human family tree, or represented two or more species of different sizes and behavior.

In a report published in the journal *Nature,* the discoverers said the skull confirmed the "taxonomic unity of *A. afarensis,*" that is, their original hypothesis that these creatures belonged to one species and not two, as other paleontologists had contended.

The discoverers described the skull as not only the youngest and largest but also the only relatively intact one of the afarensis species, which lived for almost one million years in the region from Ethiopia in the north to Tanzania in the south.

The longevity of the afarensis species was remarkable in itself, the discovery team said, noting how few detectable evolutionary changes seemed to occur between the first known afarensis specimens from 3.9 million years ago and the skull and other recently discovered fossils that are 3 million years old.

The team was headed by Dr. William Kimbel, director of paleoanthropology at the Institute of Human Origins in Berkeley, California; Dr. Donald C. Johanson, president of the institute, and Dr. Yoel Rak, a paleontologist at Tel Aviv University in Israel. Dr. Johanson was one of the discoverers of the Lucy fossils at a site about one mile away from where the skull was found, near the Awash River, and finally to discover a skull was a personal triumph for him.

Although their report emphasized the skull's scientific implications, Dr. Johanson said in an interview that its emotional effect could not be discounted.

"Until you have a skull, until you can look into one of those big orbits or hold a cranium in the cup of your hand," he said, "some people are not really satisfied that you have a new species."

Commenting on the skull's importance in an accompanying journal article, Dr. Leslie C. Aiello, a paleontologist at University College, London, said the skull and other recent findings "provide persuasive support for the idea that A. afarensis is a single, highly dimorphic species," that is, one with a large difference between men and women. She said this should "go a long way to settle some of the most heated controversies surrounding the earliest species in the human lineage."

The single-species hypothesis had been challenged by scientists who studied the striking variations in the size of afarensis fossils and decided that they were too pronounced to be included in one species. In the alternative view, larger-boned individuals represented a separate "robust" species, now extinct, which lived at the same time as the smaller species, represented by Lucy, which evolved into the Homo lineage, leading eventually to modern humans, Homo sapiens.

In this view, the two distinct lines—one leading to humans and the other to later australopithecines, a branch that became extinct one million years ago—had already diverged by three million years ago. Like many paleontologists, Dr. Kimbel's group thinks that primitive upright-walking hominids—humans and their extinct ancestors and relatives—did not diversify into discrete lineages until sometime in the half a million years after the three-million-year-old skull.

After an analysis of recent fossil discoveries, Dr. Kimbel's group attributed the size differences to sexual dimorphism. The males in the fossil record were considerably larger and heavier than the females like Lucy. The females tended to be no more than three and a half to four feet tall and weigh about 75 pounds. The males seemed to be at least five feet tall and weigh at least 110 pounds.

Each year in the Ethiopian desert, seasonal rains erode the landscape, exposing a trove of fossil bones. In 1992, exploring near their campsite, Dr. Rak and two Ethiopian assistants found the newly exposed skull scattered in more than 200 rock-encrusted fragments. The position of the fragments assured Dr. Rak that they belonged to a single individual.

In earlier research, the only fairly complete afarensis skull was a reconstruction made by Dr. Johanson and colleagues from a wide assortment of pieces from many individuals. From this paleontologists were able to make some estimates of afarensis brain sizes and other characteristics, which have generally been confirmed by the skull that has now been found.

Dr. Alan Walker, an anatomist at the Johns Hopkins University School of Medicine, who specializes in early human fossils, said it "may be stretching the point" to establish a species link between the 3.9-million-year and 3-million-year specimens solely on this frontal bone. "There's a natural temptation to try to put things together over long time periods," he said, "but we have to be cautious."

But Dr. Walker praised the richness of the new fossil site. "It's already a gold mine," he said. "Eventually these wonderful fossils will just push the debate one way or another and resolve the many issues over afarensis."

The identification of the skull as afarensis and the claim that this bolstered the single-species hypothesis was based in part on a comparison of the new bones with previous ones found both in Ethiopia and Tanzania.

Last year, Dr. Tim D. White, a paleontologist at the University of California at Berkeley, who was another of the Lucy discoverers, reported finding new fossils in the Maka area of the Awash River Valley that won more converts to the idea of sexual dimorphism in the afarensis species.

Dr. Ian Tattersall, a curator of paleontology at the American Museum of Natural History in New York City, questioned whether these specimens alone would settle the debate about afarensis being a single species. "The bones and teeth could be similar, but the appearance of these different individuals may be very different," he said. "Would they really have been able to recognize each other as members of the same species?"

Other paleontologists tended to accept the general interpretation offered by Dr. Johanson and his colleagues.

"It would now take compelling fossil evidence to start the pendulum of opinion swinging back toward the idea that there were several species" represented in the afarensis collections, Dr. Aiello said. "Stranger things, though, have been known to happen in human paleontology."

—JOHN NOBLE WILFORD, March 1994

Toolmaker's Thumb Not Unique to Humans

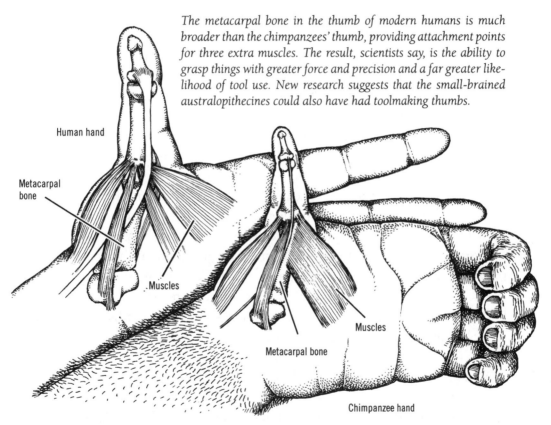

The metacarpal bone in the thumb of modern humans is much broader than the chimpanzees' thumb, providing attachment points for three extra muscles. The result, scientists say, is the ability to grasp things with greater force and precision and a far greater likelihood of tool use. New research suggests that the small-brained australopithecines could also have had toolmaking thumbs.

Human hand

Metacarpal
bone

Muscles

Muscles

Metacarpal bone

Chimpanzee hand

Patricia J. Wynne

SINCE THE FIRST stone tools appeared in Africa about two and a half million years ago, corresponding with the emergence of the earliest direct human ancestors, anthropologists have long assumed that the ability to make tools

211

was somehow linked to brain size. The earliest members of the genus *Homo,* with relatively large brains, had the ability, so it was thought, but not their smaller-brained primate relatives.

This standard assumption is now being challenged by an anatomical study that has produced a rule of thumb that is more than a figure of speech.

Applying the rule with surprising results, a scientist has determined that all hominids, small-brained australopithecines as well as large-brained *Homo*s, could have been using tools at least two million years ago. The new findings provide further evidence to discredit a theory, once widely held, that no two species of hominids with similar cultural abilities, like tool-making, could possibly have existed at one time and place.

In a close examination of the modern human thumb, Dr. Randall L. Susman, an anatomist at the State University of New York at Stony Brook, got the idea for the study. He identified a particular bone whose size and shape are related to the thumb muscles that enable people to grasp things with force and precision. This bone, the first metacarpal, which is embed-ded in the palm of the hand at the base of the thumb, is broader in humans than in chimpanzees, the closest living relative of humans, because it must support three extra power muscles, which chimpanzees lack.

Like humans, chimpanzees can oppose their thumbs to the rest of their fingers and so can use sticks for digging and stones for simple tasks. But their hands are better adapted for swinging from branches. Only humans can apply the significant force needed for opening jar lids and grasping ham-mers or the dexterity for holding a pencil, needle or surgeon's scalpel.

Once Dr. Susman recognized the revealing nature of the first metacarpal bone, he reasoned that a comparative analysis of fossil thumbs should be a reliable way of determining which of the early hominids would have had hands that functioned in a way similar to those of modern humans. The results of the study were reported in the journal *Science.*

Of the four fossil species examined, only *Australopithecus afarensis,* the species associated with the famous Lucy skeleton, failed the rule of thumb. Its first metacarpal bone was narrow and ape-like, incapable of supporting the enhanced thumb musculature required for toolmaking. This was to be expected, because this species lived four million to three million years ago, long before any evidence for toolmaking. Another member of this group,

Australopithecus africanus, was not tested, but it became extinct two million years ago and tools have never been found with its fossils.

Nor was it surprising that the thumbs of *Homo erectus* and *Homo sapiens neanderthalensis* were of the toolmaking and tool-using type. *Homo erectus* evolved some 1.8 million years ago and is the immediate ancestor to modern humans. The Neanderthals, which became extinct less than 40,000 years ago, were apparently a subspecies of modern humans with more in common than not. Ample evidence of stone and bone implements has been found at sites associated with both groups.

A result that may be harder for paleoanthropologists to accept is the discovery through this analysis that *Paranthropus robustus* also had a toolmaker's thumb. This *Paranthropus,* also known as *Australopithecus robustus,* emerged at least two and a half million years ago and was a longtime contemporary of *Homo erectus.* But it looked more like a gorilla, had a brain size roughly half that of *Homo erectus* and one third that of *Homo sapiens* and belonged to a branch of the hominid family tree that became extinct about a million years ago.

The result did not surprise Dr. Susman, who for several years has contended that some of the fossil bones associated with tool discoveries in South Africa belonged not to *Homo erectus* but to *Paranthropus robustus.* The thumb tests, he reported, "indicate that tools were likely to have been used by all early hominids at around two million years ago."

The technique, he said, should lend itself to wide application. An advantage is that it relies on a single thumb bone, not a complete hand, and one that is well represented in the fossil record.

In comments accompanying the published report, Dr. Leslie C. Aiello, a paleontologist at University College, London, praised the analysis as "simple, but elegant." But she cautioned against making any sweeping inferences based on the research until it can be more firmly established that the fossils tested were from a *Paranthropus* and not a *Homo erectus.*

"Two ambiguous bones are insufficient evidence upon which to determine whether more than one contemporaneous species of our early ancestors made and used tools," Dr. Aiello wrote.

Dr. Susman suggested in an interview that Dr. Aiello must not have been aware that more recent research had, he said, cleared up the ambiguity and identified these fossils from Swartkrans Cave in South Africa as definitely belonging to *Paranthropus.*

Other scientists reserved judgment, but foresaw controversy. Even though the single-species hypothesis is dead, said Dr. Pat Shipman, a paleontologist at the Johns Hopkins University School of Medicine, some scientists in the field are still resistant to the idea that more than one hominid could have been making tools at the same time and place.

If they eventually award a thumbs-up to the research, scientists will face other questions of interpretation. How could two species manage to have developed similar toolmaking abilities? How did they co-exist? Why did one line of toolmakers survive to the present, while the other fell by the wayside?

Five years ago, when Dr. Susman first offered indirect evidence for *Paranthropus* toolmaking but before he hit upon the thumb test, Dr. Shipman suggested an explanation for why two different species might have made the same type of simple tools, mostly choppers, scrapers, sharp flakes and spheroids for cracking nuts or bones.

Making these tools might have been fairly simple, she said, but the idea of turning a blunt, heavy rock into something small and sharp was an inspired conceptual leap. Perhaps the idea first came to the larger-brained *Homo,* but the practice was eventually imitated by the contemporary species of more limited intellect.

The early tools, being so unspecialized, were equally good for many tasks. "The same spheroid that broke open nuts for one species might have cracked open marrow-bearing bones for another," Dr. Shipman wrote.

Differences in eating habits and other behavior could have held competition to a minimum and ensured a long co-existence between the two species in the same geographical area, paleontologists said. *Paranthropus* was a vegetarian and *Homo* an omnivore.

The apparent toolmaking skill will deprive scientists of one item on the list of reasons for the extinction of *Paranthropus,* the last of the australopithecines. When it occurred more than a million years ago, paleontologists thought, climate change was sweeping Africa and drier conditions led to a loss of forest habitats. Perhaps the vegetarian *Paranthropus* could not adapt to change quickly enough, but *Homo erectus* could. The new evidence seems to show that *Paranthropus* had more talents than given credit for.

The new thumb test, Dr. Susman said, might also help clear up questions about who produced three stone tools found in 1960 in Olduvai Gorge, in Tanzania. The discovery by Mary Leakey of a partial hand of *Homo habilis,* the earliest identified member of the human lineage, prompted the inferences that early *Homo* was the first toolmaker. Some of those tools could have been made by *Homo*'s vegetarian neighbors.

—JOHN NOBLE WILFORD, September 1994

Human Ancestors' Earliest Tools
Found in Africa

STONE TOOLS found in the desiccated fossil fields of Ethiopia have been dated at 2.6 million years old, making them the earliest known artifacts created by humans or their direct ancestors.

Simple cobbles fractured and flaked for use in scraping and cutting, the tools were made and gripped by hands apparently well evolved toward the hands of humans and conceived by minds somewhat less far along the evolutionary path. These were toolmakers, though, probably adjusting to new environmental conditions, a cooler climate and changing vegetation, by trying out new means of getting and processing food. Whoever they were, whatever their needs, they certainly were innovators back near the beginning of the human experience.

The discoveries provided strong evidence that toolmaking, once thought to be a skill special to the *Homo* genus, began long before its emergence, perhaps by half a million years. Fossils of the early human ancestors themselves have yet to be identified in the tool-bearing sediments of the Ethiopian site. But the scientists reporting the new data said it was a reasonable assumption that the tools could have been made by more primitive members of the human lineage, the australopithecines, the only hominids known to have existed at that time.

The existence of some early stone tools in the ground near the Gona River had been known since the 1970's. Preliminary analysis indicated that they were probably much older than any previously discovered artifacts, but nothing was firm. When new explorations in the Gona region over three years uncovered thousands of these crude tools, more precise geological dating tests were conducted.

Sileshi Semaw, an Ethiopian archaeologist who is a doctoral student at

Rutgers University in New Brunswick, New Jersey, announced the test results at a recent meeting of the American Association of Physical Anthropology in Oakland, California. Argon isotopic analysis and other evidence, he reported, showed ages of 2.5 million to 2.6 million years for the sediments where the tools were found, with the true date probably being closer to the greater age.

In that case, Mr. Semaw concluded, this collection of artifacts is the oldest known in Africa and thus the world. Tools discovered at Olduvai Gorge in Tanzania by Louis and Mary Leakey, the famous husband-and-wife team of fossil hunters, are about two million years old, and other recent finds in Kenya and Ethiopia have been determined to be several hundred thousand years older, but not as ancient as the Gona artifacts.

Other archaeologists and paleontologists agreed that the findings were an important step in the study of human origins, especially in the poorly understood period from three million to two million years ago. Dr. Ian Tattersall, a paleoanthropologist at the American Museum of Natural History in New York and author of *The Fossil Trail*, published by Oxford University Press, said, "The new set of dates certainly firm up dates for very early toolmaking, and that's certainly a significant finding."

But at the scientific conference, the stone tools of Gona were overshadowed by rancorous charges and countercharges touched off by Mr. Semaw's accusations of "a violation of professional ethics" by another research group that he said had invaded his site last October. The group, from the Institute of Human Origins in Berkeley, California, headed by Dr. Donald C. Johanson, discoverer of the celebrated Lucy fossils, flatly denied the allegations.

The tempest has yet to subside, however, as scientists in the often-contentious field of paleoanthropology are taking sides and airing bitter complaints about claim-jumping.

Dr. J. W. K. Harris, who has explored the Ethiopian fossil beds for years and is now chairman of the Rutgers anthropology department, was among the first to find tools in the Gona region, where he discovered a few scattered samples in 1976 and 1977. Beginning in 1992, he and Mr. Semaw conducted the first systematic surveys there, identifying 21 sites containing thousands of stone tools and ancient animal remains. Most of the tools were stone pebbles, usually of lava, that had been struck repeatedly with a second pebble as a hammer. This produced a crude chopper or scraper, as well as numerous sharp flakes for cutting.

Dr. Craig S. Feibel, a Rutgers geologist, collected samples of volcanic ash for use in dating the material. Then Dr. Paul R. Renne, president of the Berkeley Geochronology Center in California, ran tests with a recently perfected geological dating technology known as argon-argon single crystal laser fusion. He has also applied the technology in dating the crater in the Yucatán Peninsula of Mexico that is implicated in the asteroid collisions that may have wiped out the dinosaurs. Previous techniques required much larger samples than are usually available.

"It's the most precise, most broadly applicable test out there in archaeology and geology," Dr. Renne said.

The analysis produced a date of 2.517 million years, with only a slight margin of error. Because the dated ash lay just above the 15-foot-thick tool sediments, Dr. Renne said, the tools are assumed to be somewhat older. By how much? Rock found below the sediment held the record of a previously dated reversal of Earth's magnetic field. The date for that is about 2.6 million years ago, and the artifacts appeared to be much closer to that boundary than to the volcanic ash that is overlying the sediments.

"Now everybody wants to find the toolmaker," Dr. Renne remarked. The research team from the Institute of Human Origins in Berkeley, which Mr. Semaw accused of trespassing on his site, recovered some fossil teeth there in October. Officials of the institute said the fossils have not been analyzed and so it is not known if they are associated in time and place with the tools.

Dr. Philip Rightmire, a paleoanthropologist at the State University of New York at Binghamton, said that it was highly unlikely that the tools were made by a *Homo* species, for which there is no fossil evidence any earlier than two million years ago. And the date is too recent for the toolmaker to be *Australopithecus afarensis*, the family Lucy belonged to, which seems to disappear in the fossil record after 2.9 million years ago.

One possibility, Dr. Rightmire said, is that the Gona toolmakers were *Australopithecus aethiopicus*, a little-known but more robust species than the afarensis family.

Although the direct human lineage evolved from australopithecines, other species of this branch of the family tree continued on their separate ways until about one million years ago, when all traces of australopithecines vanish.

Dr. Harris and Mr. Semaw said the new findings could help scientists assess rival hypotheses proposed for the beginning of tool manufacturing. The standard hypothesis, now called into question, associated toolmaking with the first signs of significant brain expansion in the first *Homo* species, which emerged about two million years ago and was accordingly given the name *Homo habilis,* "handy man."

Other paleontologists contend that when early human ancestors began walking upright, their hands were freed for the manipulation and use of tools. So it is not inconceivable that the earlier australopithecines could have been using tools even before 2.6 million years ago. Indeed, Dr. Robert Foley, a paleontologist at Cambridge University in England, has argued, on the basis of chimpanzees' use of wooden sticks to collect termites and of stones to crack nuts, that simple tool use is a much earlier trait, perhaps reaching back before the divergence of the pre-human lineage from the African apes.

In any event, the simple tool technology of Gona and the Olduvai Gorge remained little changed for a million years, until after the appearance of the more advanced *Homo erectus* 1.7 million years ago. The new species made more complex and varied tools, including the first hand axes.

The stone tools of Gona, paleontologists said, promise to have a more direct bearing on questions of early human behavior in the critical, but poorly studied, period around 2.5 million years ago. Several lines of evidence indicate that a cooler climate swept across Africa at that time. Stone tools, Dr. Harris said, were probably "a response to a different environment, dietary changes and competition with other hominids—a matter of survival."

—JOHN NOBLE WILFORD, April 1995

Bones in China Put New Light on Old Humans

DISCOVERIES OF BONES and stone tools in a cave in central China have provided compelling evidence that human ancestors began their global migrations out of Africa much earlier than once thought—at least 1.9 million years ago.

Digging in clay and gravel deposits in the cave, Chinese paleontologists came upon a fragment of lower jaw with a molar and adjacent pre-molar and also an isolated incisor. The teeth resembled those of extremely early human ancestors not previously known to have reached Asia. The same sediments yielded a spherical cobble hammerstone and a large battered flake that appeared to be the type of tools associated with especially primitive human forerunners.

Other scientists who have examined the material called the discovery surprising, exciting and provocative. The evidence could overturn current thinking that *Homo erectus,* the immediate ancestor of *Homo sapiens,* was the first species in the human family to pick up and leave Africa. It may instead have been a more primitive species resembling *Homo habilis* or *Homo ergaster.*

And it just may be, some scientists say, that this early species evolved into *Homo erectus* in Asia, not Africa, suggesting a more intercontinental character to the early history of human origins. No one is saying that the findings dispute the place of Africa as the cradle of humanity, where the first hominids—the genus *Homo* and other close relatives—emerged after splitting off the ape lineage. Nor does the discovery necessarily conflict with a growing consensus that modern *Homo sapiens* probably emerged in Africa no more than 200,000 years ago.

But until recently, there had been little unambiguous evidence that hominids had reached eastern Asia any earlier than one million years ago. In 1991, a small jaw and some tools, dated at 1.8 million years ago, were found at Dmanisi, Georgia, a surprising sign of early movement toward Asia. In 1994 scientists announced a similarly early date for a juvenile *Homo erectus* from Indonesia, though there have been doubts about the reliability of the dating.

The new specimens were found at Longgupo Cave in the late 1980's by Dr. Huang Wanpo and Dr. Gu Yumin, paleontologists at the Institute of Vertebrate Paleontology and Paleoanthropology in Beijing. The limestone cave is 12 miles south of the Three Gorges area of the Yangtze River in eastern Sichuan Province.

Among the hominid bones the scientists found fossils of extinct species of mastodon, horse and pygmy giant panda. These indicated an age of 1.5 million to 2 million years for the material. But no one could be sure.

An international team of paleontologists, archaeologists and geologists has since analyzed and dated the fossils and tools. The team was organized by Dr. Russell L. Ciochon, a paleontologist at the University of Iowa in Iowa City, and Dr. Roy Larick, an archaeologist at the University of Massachusetts in Amherst.

In the first definitive report by the team, published in the journal *Nature,* Dr. Huang, Dr. Ciochon and their colleagues said three different dating methods were used to establish and confirm the 1.9-million-year age of the fossils. They also noted that the teeth were "demonstrably more primitive than that seen in Asian *Homo erectus.*"

The scientists thus concluded: "The new evidence suggests that hominids entered Asia before two million years ago, coincident with the earliest diversification of genus *Homo* in Africa. Clearly, the first hominid to arrive in Asia was a species other than true *Homo erectus,* and one that possessed a stone-based technology."

Commenting on the report, Dr. Bernard Wood and Dr. Alan Turner, paleoanthropologists at the University of Liverpool in England, wrote in another article in *Nature* that the discovery "is of major importance" and lends support to "other, less well-substantiated claims that hominids traveled even further, and occupied China some 1.9 million years ago."

The two British scientists noted that the remains of the Longgupo hominid are meager, but agreed that they were sufficient to show that the hominids in the cave were much more primitive than *Homo erectus*.

The isolated incisor, for example, was almost indistinguishable in size and shape from teeth known for *Homo habilis,* a small tool-using species associated with Olduvai Gorge in Tanzania and Lake Turkana in Kenya. The crudely worked cobbles from Longgupo Cave were similar to those excavated at Olduvai and presumably made by *Homo habilis.* The cusps of the molar from the cave also resembled teeth of habilis.

But the double root of the pre-molar was characteristic of *Homo ergaster,* a larger hominid from Kenya. Some fossils of habilis and ergaster have been dated as old as 2.2 million years. The earliest recognizable member of the *Homo* genus, currently known as *Homo rudolfensis,* appeared more than 2.5 million years ago in the Rift Valley of East Africa.

Dr. F. Clark Howell, a paleontologist at the University of California at Berkeley, who is familiar with the specimens, said he agreed with the interpretation that the Longgupo hominid was not erectus but probably ergaster. "This definitely has evolutionary implications," he said.

But Dr. Howell withheld endorsement of what may be the most sensitive aspect of the new report. If there were hominids in Asia before erectus and erectus appeared there soon, Dr. Ciochon explained, "*Homo erectus* may actually have evolved within Asia from an ancestor like the hominids found at Longgupo."

Dr. Wood and Dr. Turner conceded that the findings implied the possibility that erectus evolved in Asia and then spread back to Europe and Africa. But Dr. Howell resisted. "I want to see the last *t* crossed before I use the material to hang new hypotheses on," he said in a telephone interview.

Dr. Ciochon elaborated on this back-migration thesis in an interview and in an article in *Natural History* magazine. It is possible, he said, that *Homo erectus* and this early hominid in China—which he tentatively identified as *Homo habilis*—were sister species that spread out of Africa in successive migrations.

But an alternative interpretation could not be overlooked. *Homo erectus,* he said, "may have evolved in Asia from the species represented at Longgupo." Some paleontologists have for some years suspected that *Homo erectus* is an Asian species. The first known specimens of that species, the famous Java man and Peking man, were discovered long before any comparable fossils were excavated in Africa. The Java and Peking fossils accounted in part for the widespread belief early in this century that humans originated in Asia, though Darwin had correctly predicted that Africa would be shown to be the ancestral home.

As the discovery at Longgupo Cave reveals, the dispersal of human ancestors out of Africa is proving to be more ancient and surprising than anyone suspected just a few years ago.

—JOHN NOBLE WILFORD, November 1995

Neanderthals' stone tools remained unchanged for thousands of years, like typical point below. Tools of early moderns in Europe were more specialized, like those below from southwest France and nearby areas in Spain, contributing to theory that more advanced moderns soon bested Neanderthals in competition for resources. This theory holds that early moderns were able to dominate critical advantages of valleys–water, shelter, edible plants and game–and eventually to expel Neanderthals from these hospitable territories to harsher high altitudes with fewer supplies.

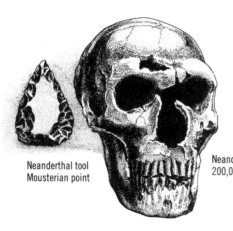

Neanderthal tool
Mousterian point

Neanderthal skull
200,000 to 35,000 years ago

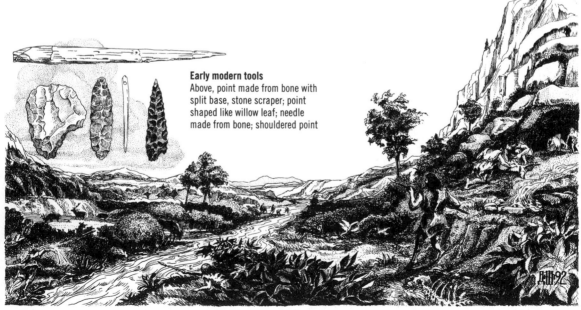

Early modern tools
Above, point made from bone with split base, stone scraper; point shaped like willow leaf; needle made from bone; shouldered point

Dimitry Schidlovsky

Neanderthals and Modern Humans Co-existed Longer Than Thought

AS FAMILIAR as Neanderthals may seem, with their robust bodies, stooped shoulders and beetle brows, the very image of the extinct cave dwellers of popular lore, their place in human evolution and their fate remain deep mysteries.

Like modern humans, Neanderthals were presumably descendants of archaic *Homo sapiens,* but evolved into a distinct branch of the human family over several hundred thousand years, living in relative isolation in Europe and parts of the Middle East. Their brains were as large as those of modern humans, if not a little larger. But they appeared to be deficient in tool technology, art and other innovative turns of the mind.

So they were no match for the taller, more slender modern humans who probably migrated from Africa through the Middle East some 120,000 years ago, proceeded into southeastern Europe and western Asia and eventually arrived in western Europe about 40,000 years ago. These clever newcomers, usually known as Cro-Magnons but in every respect physically like people today, soon replaced Neanderthals in their European homeland, driving them to extinction, but apparently not as swiftly as previously thought.

New fossil evidence shows that the enigmatic Neanderthals, the last competitors to modern humans in their ascent to global dominion, were still living in Croatia as recently as 33,000 years ago and in southern Spain only 30,000 years ago, about 6,000 years after it was assumed they had died out.

Anthropologists said the new findings might not resolve the many questions about the origin of Neanderthals, their exact place on the family tree or the reasons they lost out in the competition with modern human beings. But the research has established that Neanderthals and modern *Homo sapiens* co-existed in western Europe for at least 10,000 years and, scientists said, could have shared habitats in what must have been a complex relationship, possibly involving some interbreeding as well as relentless competition. Until now, the last evidence for a Neanderthal was 36,000 years ago, at an archaeological site near the village of St. Cesaire in France.

"The existence of Neanderthal populations in southern Spain long after modern humans arrived in the north suggests that Neanderthals were not quickly replaced due to the overwhelming superiority of modern humans, as many archaeologists have contended," Dr. Jean-Jacques Hublin, director of research at the Museum of Man in Paris, said in reporting the dating of fossil bones from a Neanderthal site in Spain.

Dr. Hublin and a team of Spanish archaeologists and French dating specialists made their discovery at Zafarraya Cave near the Mediterranean coastal city of Málaga. They uncovered a well-preserved jaw with characteristics typical of Neanderthals, stone tools of the Mousterian style associated with Neanderthals and the teeth of an ibex, perhaps the prey of Neanderthal hunters or scavengers. Radiocarbon and thorium-uranium dating tests on the bones and teeth, combined with an analysis of the tool technology, led to the conclusion that Neanderthals had been living there as recently as 30,000 years ago.

The results of the dating tests are reported in *Archaeology*, a magazine of the Archaeological Institute of America. The Transactions of the Academy of Sciences of Paris is publishing a more detailed description of the work by Dr. Hublin and other explorers of Zafarraya Cave, including two Spanish archaeologists, Dr. Cecilio Barroso Ruiz and Dr. Paqui Medina Lara.

Dr. Erik Trinkaus, an anthropologist at the University of New Mexico in Albuquerque and author of several books on Neanderthals, said the new date was based on careful research and established "the last Neanderthals that we know of." If any more recent fossils should turn up, they would probably also be in Spain or Portugal, he said, because these were "geographical cul-de-sacs" where Neanderthals must have survived longer, out of the way of the intrusive modern humans.

In the rest of Europe by this time, Dr. Trinkaus noted, modern humans had spread from east to west; there is no evidence that people could have crossed from Africa into Europe by way of Gibraltar. Though the extent of mixing between the modern and Neanderthal populations is not known, Dr.

Trinkaus said that the disappearance of Neanderthals could be attributed in part to interbreeding as well as competition.

Once it was suspected that Neanderthals and modern *Homo sapiens* were sufficiently distinct species and thus could not have interbred—or as the anthropologists often put it, "exchanged genes." Now anthropologists lean to the view that the relationships between the two human groups might have been close enough for some interbreeding, perhaps like the kinship of dogs and wolves today.

Dr. Fred H. Smith, a paleontologist at Northern Illinois University in DeKalb who is an authority on Neanderthal history, said the new date for the last Neanderthals was not too surprising because other discoveries had been pointing to similar pockets of populations holding out against extinction. At the Vindija site in Croatia, Dr. Smith said, he recently dated Neanderthal fossils at 33,000 years old and established that the two human populations were living at the same time in that region.

Recent excavations in Israel suggest that Neanderthals and modern humans might have overlapped there as early as 100,000 years ago. Their stone tools and bones have been found in caves only a few miles apart. Whether the two human groups were exact contemporaries in the region, or moved in and out without meeting, has not been determined. Nor is it possible to determine what their relations were with each other, if they did share the place at the same time.

Dr. Smith said that such discoveries, along with the recent date for the Zafarraya fossils, are contributing to the ferment in Neanderthal studies.

"Things are getting exciting again," he said.

—JOHN NOBLE WILFORD, August 1995

Fossils Called Limb
in Human Family Tree

SPANISH PALEONTOLOGISTS say 800,000-year-old fossils they found in a limestone cave represent an entirely new species of human ancestor and could be the last common ancestor of modern humans and their extinct cousins, the Neanderthals.

Other scientists are not so sure. They agree that the fossils, whatever their species, are remains of the earliest known Europeans. As such, the fossils are critical to an understanding of how Europe was first settled by hominids, the family of all human species, alive and extinct. But scientists expressed serious reservations about assigning the Spanish fossils to a separate limb on the family tree.

A Spanish team of paleontologists, led by Dr. José M. Bermúdez de Castro of the National Museum of Natural Sciences and Dr. Juan Luis Arsuaga of the Complutense University, both in Madrid, based their case for a new hominid species on the bones of a boy with a remarkably modern face. The partial remains of the boy, about 10 or 11 years old, and five other individuals were discovered in 1995 at the Gran Dolina cave in the Atapuerca hills near Burgos, Spain.

Everything about the boy's cranium, lower jaw and teeth, the researchers could see, was primitive. But between the brow and the jaw, the sunken cheeks, projecting nose and other traits suggested "a completely modern pattern of midfacial topography," the paleontologists wrote in a report published in the journal *Science*.

This combination of modern and primitive characteristics seems to set the Atapuerca boy apart from any previously recognized species, the researchers said. No hominids had been known to have developed such a modern face earlier than 200,000 years ago. And the primitive aspects do not seem to be like those of *Homo heidelbergensis,* the Heidelberg man of 500,000 years ago, which had been the earliest known hominid in western Europe.

Accordingly, the paleontologists decided the fossils belonged to a separate species and proposed that it be named *Homo antecessor,* meaning "one who goes before."

As if proposing a new hominid species was not enough to provoke impassioned debate, the Spanish team invited further controversy by identifying the 800,000-year-old species, not *Homo heidelbergensis* or the even earlier *Homo erectus,* as the probable common ancestor of Neanderthals and modern humans, *Homo sapiens.*

The journal quoted Dr. Antonio Rosas, another paleontologist from the Madrid museum and member of the team, as saying the boy's facial traits are "exactly the morphology we would imagine in the common ancestor of modern humans and Neanderthals."

One way to account for *Homo antecessor*'s role as ancestor to both Neanderthals and modern humans, the Spanish researchers said, is to postulate the origin of the new species in Africa, though no similar fossils have been found there. Some members of *Homo antecessor* in Africa gave rise eventually to *Homo sapiens,* according to the hypothesis, while others migrated to Europe, possibly about a million years ago. The European branch then evolved into *Homo heidelbergensis,* which in turn led to Neanderthals.

But many paleontologists are not yet prepared to accept *Homo antecessor* as a distinct species.

"I'm reluctant to endorse this new species," said Dr. Philip Rightmire, a paleoanthropologist at the State University of New York at Binghamton. "I wonder if the facial characteristics of one juvenile are really diagnostic. It's tricky to compare children to adults and on that basis establish a new species."

Dr. Fred H. Smith, a paleoanthropologist at Northern Illinois University in DeKalb who specializes in European hominids, said the Gran Dolina fossils were "interesting and important" but "insufficient evidence for a new species."

Dr. Smith said the fossils would probably turn out to be either an early *Homo heidelbergensis* or late *Homo erectus.* Dr. Rightmire said he saw no reason yet to reconsider his idea that *Homo heidelbergensis* was the last common ancestor of Neanderthals and modern humans.

But Dr. Rosas said the few pieces of facial bones from the adults collected in the cave, a discovery announced last in 1995, also showed some of the same modern-looking characteristics found in the boy's face. He insisted that the team had enough evidence to support the designation of a new species, while acknowledging that "people are probably going to need some time to accommodate this proposal."

—JOHN NOBLE WILFORD, May 1997

Three Human Species Co-existed Eons Ago, New Data Suggest

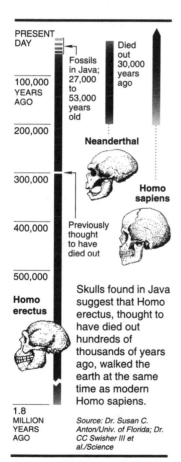

PRESENT DAY

100,000 YEARS AGO

200,000

Fossils in Java; 27,000 to 53,000 years old

Died out 30,000 years ago

Neanderthal

300,000

Homo sapiens

400,000

Previously thought to have died out

500,000

Homo erectus

Skulls found in Java suggest that Homo erectus, thought to have died out hundreds of thousands of years ago, walked the earth at the same time as modern Homo sapiens.

1.8 MILLION YEARS AGO

Source: Dr. Susan C. Anton/Univ. of Florida; Dr. CC Swisher III et al./Science

The New York Times

Skulls found in Java suggest that Homo erectus, *thought to have died out hundreds of thousands of years ago, walked the Earth the same time as modern* Homo sapiens.

SCIENTISTS have found stunning new data showing that a third human species apparently co-existed on Earth with two others as recently as 30,000 years ago.

In research that could redraw the human family tree and is certain to be controversial, the scientists reexamined two major fossil sites along the Solo River in Java and found that an early human relative, *Homo erectus,* appeared to have lived there until about 53,000 to 27,000 years ago.

Writing in the journal *Science,* the scientists said the new dates were "surprisingly young and, if proven correct, imply that *H. erectus* persisted much longer in Southeast Asia than elsewhere in the world."

Confirmation of the new dates would mean that at least in Java, this archaic species, which evolved 1.8 million years ago, survived some 250,000 years after it was thought to have become extinct. This surviving population of *H. erectus* in Indonesia would have been alive at the same time as anatomically modern humans—*Homo sapiens*—and also Neanderthals.

231

The Neanderthals, which lived in Europe and western Asia for some 300,000 years, appear to have made their last stand 30,000 years ago in southern Spain. By then, modern *H. sapiens,* who are widely thought to have evolved in Africa 200,000 to 100,000 years ago, had spread all over Africa and Eurasia, as far as Australia. It is not known how much contact the three species had.

In any case, specialists in human evolution noted, the new findings suggest that the present phenomenon of a solitary human species on Earth may be more the exception than the rule. Until about a couple of decades ago, scientists conceived of the human lineage as a neat progression of one species to the next and generally thought it impossible that two species could have overlapped place or time.

Another implication of the more recent date for *H. erectus* is to undercut a pillar of the multiregional theory for the origin of modern *H. sapiens.*

As the most advanced known representatives of *H. erectus,* the Java fossils have appeared to be a clear intermediate step in the evolution of *H. erectus* in Southeast Asia to the first Australians, who were modern *H. sapiens.* This has lent support to the idea that modern humans emerged gradually out of *H. erectus* in many parts of the world. The alternative and more favored out-of-Africa theory holds that modern humans evolved in Africa less than 200,000 years ago and displaced *H. erectus* as they migrated to the ends of the Earth.

The team of scientists, led by Dr. Carl C. Swisher, 3d, of the Berkeley Geochronology Center in California, concluded that it was "no longer chronologically plausible" to argue that the Java *H. erectus* evolved into Asian *H. sapiens.* From earlier fossil evidence, Australian *H. sapiens* are at least 30,000 years old, and could be much older, judging by rock art.

"The multiregionalists will have to do some fast talking to explain this," said Dr. Philip Rightmire, a paleoanthropologist at the State University of New York at Binghamton. "It's quite a blow for them to absorb. But neither side has won the day yet in this theoretical battle."

Dr. Milford Wolpoff, a paleoanthropologist at the University of Michigan who is an outspoken leader of the multiregional theorists, questioned both the accuracy of the dates and the identification of the skulls at the Java sites, contending that they were *H. sapiens* and not *H. erectus.* Dr. Wolpoff said these questions should have been answered more convincingly before the team published its report.

As one who has studied the skulls at Ngandong, one of the two sites, and compared them with early Australian *H. sapiens,* Dr. Wolpoff said the idea of an ancestral "link between them is incontrovertible."

In an accompanying article in *Science,* Dr. Alan Thorne of the Australian National University in Canberra, one of Dr. Wolpoff's allies, said, "There is a great long list of characters that are the same in the Solo skulls and the earliest known human people from Australia."

Even if the Java fossils are indeed relatively young, Dr. Thorne added, they look so much like the Australian fossils that the two species may have shared a recent ancestor.

Both Dr. Rightmire, an authority on *H. erectus,* and Dr. Susan C. Anton, a paleoanthropologist at the University of Florida who was a member of Dr. Swisher's team, said they were satisfied that the Java specimens were *H. erectus,* though the skulls did show signs of their having evolved a somewhat larger brain than earlier members of the species. That *H. erectus* and *H. sapiens* now appear to have overlapped, Dr. Anton said, "raises the possibility of gene flow between the two lines."

The Java fossils were discovered in the 1930's by Dutch geologists. Over the years, various efforts to date the specimens have yielded ages of as high as 300,000 and 250,000 years and, recently, as low as 100,000 years.

The most reliable means of dating such ancient material is to determine the age of volcanic ash in the sediments, but none is associated with the Solo specimens. Similarly, a number of other techniques were not suitable, and the skulls themselves could not be dated because they would have been damaged.

Dr. Swisher explained in an interview that he had dug a test trench at the Ngandong site, on a terrace at the bend in the Solo River. He noted that the layer of sediment where the skulls had been found also contained teeth of water buffalo. The teeth were analyzed through two techniques for measuring the radioactive decay of uranium that the teeth had absorbed from the soil. This indicated how long the teeth had been buried.

To assure themselves that they were sampling the same sediment layer where the *H. erectus* skulls had been excavated, the scientists said, they compared and dated similar water buffalo teeth collected by the Dutch geologists at the time of the original discoveries. The specimens are kept at Gadjah Mada University in Yogyakarta, Indonesia. The dating analysis was conducted at McMaster University in Hamilton, Ontario.

Some scientists said they were still concerned that erosion and river currents could have mixed up older *H. erectus* skulls with younger water buffalo teeth, but Dr. Swisher said this was unlikely to have occurred in the same way for 12 different skulls at two widely separated sites.

If the dates are right, as Dr. Swisher's team noted at the conclusion of its report, "the temporal and spatial overlap between *H. erectus* and *H. sapiens* in Southeast Asia is reminiscent of the overlap of Neanderthals (*H. neanderthalensis*) and anatomically modern humans in Europe."

—JOHN NOBLE WILFORD, December 1996

Neanderthal DNA Sheds New Light on Human Origins

A HAUNTINGLY BRIEF but significant message extracted from the bones of a Neanderthal that lived at least 30,000 years ago has cast new light both on the origin of humans and Neanderthals and on the long-disputed relationship between the two.

The message consists of a short strip of the genetic material DNA that has been retrieved and deciphered despite the age of the specimen. It indicates that Neanderthals did not interbreed with the modern humans who started to supplant them from their ancient homes about 50,000 years ago.

The message also suggests, said the biologists who analyzed it, that the Neanderthal lineage is four times older than the human lineage, meaning that Neanderthals split off much earlier from the hominid line than did humans.

The finding, made by a team of scientists led by Dr. Svante Paabo of the University of Munich in Germany, marks the first time that decodable DNA has been extracted from Neanderthal remains and is the oldest hominid DNA so far retrieved. The DNA was extracted from the original specimen of Neanderthals, found in the Neander Valley near Düsseldorf, Germany, in 1856, and is now in the Rheinisches Landesmuseum in Bonn.

"This is obviously a fantastic achievement," said Dr. Chris Stringer, an expert on Neanderthals at the Museum of Natural History in London.

Many anthropologists had tried to extract DNA from Neanderthal bones without success. "Clearly, it's a coup," Dr. Maryellen Ruvolo, an anthropologist at Harvard University, said of the Munich team.

The Neanderthals were large, thick-boned individuals with heavy brows and a braincase as large as that of modern humans but stacked behind the face instead of on top of it. They lived in Europe and western Asia from 300,000 years ago, dying out about 270,000 years later.

For the latter part of that period they clearly co-existed with modern humans but the relationship between the two groups, whether fraternal or genocidal, has been debated ever since the first Neanderthal was discovered. Early humans and Neanderthals may have interbred, as some scientists contend, with modern Europeans being descended from both; or the two hominid lines may have remained distinct, with humans displacing and probably slaughtering their rivals.

The new finding, reported in the journal *Cell*, comes down firmly on the side of Neanderthals having been a distinct species that contributed nothing to the modern human gene pool. Calling it "an incredible breakthrough in studies of human evolution," Dr. Stringer said the results showed Neanderthals "diverged away from our line quite early on, and this reinforces the idea that they are a separate species from modern humans."

The new work was also praised by scientists who study ancient DNA, a lively new field, which has included reports of DNA millions of years old being retrieved from dinosaur bones, fossil magnolia leaves and insects entombed in amber. Although these reports have appeared in leading scientific journals like *Science* and *Nature*, other scientists have been unable to reproduce them. In at least one case the supposed fossil DNA was contaminated by contemporary human DNA.

But a leading critic of these claims of ancient DNA extraction, Tomas Lindahl of the Imperial Cancer Research Fund in England, has given the new work his seal of approval, calling it "arguably the greatest achievement so far in the field of ancient DNA research." The Munich team took great pains to verify that it had a genuine sample of Neanderthal DNA. Working in sterile conditions, team members isolated it in two different laboratories and distinguished it from the human DNA, which contaminated the bones. Their work is "compelling and convincing," Dr. Lindahl wrote in a commentary in *Cell*.

The DNA recovered from the Neanderthal is known as mitochondrial DNA, a type especially useful for monitoring human evolution. Mitochondria are tiny, bacteria-like organelles within the cell and possess their own DNA. They exist in eggs, but not in sperm, and so are passed down through the female line. Unlike the main human genes on the chromosomes, which get shuffled each generation, the only change to mitochondrial DNA is the accidental change caused by copying errors, radiation or other mishaps.

Once a change, or mutation, becomes established in mitochondrial DNA, it gets passed on to all of that woman's descendants. Tracking mutations is a powerful way of constructing family trees. The branch points on such a family tree can also be dated with some plausibility if at least one of them can be matched to a known event in the fossil record, like the parting of the human and chimpanzee lines.

The Munich team focused on a particularly variable region of mitochondrial DNA and reconstructed the Neanderthal version of it, 378 units in length. Comparing it with modern human DNA from five continents, they found it differed almost equally from all of them, signaling no special relationship with contemporary Europeans, as wold have been expected if Neanderthals and modern humans interbred.

In addition, the family tree of Neanderthal mutations, when compared with those of the chimpanzee and human, yielded a distinctive pattern of variations. In the authors' interpretations, Neanderthals branched off the hominid line first, followed by humans much later.

According to the fossil and archaeological record, humans and Neanderthals diverged at least 300,000 years ago. The mitochondrial DNA evidence agrees well with this date, the authors say, since individual genes would be expected to diverge before the divergence of populations.

From fossil evidence the human and chimpanzee lines are thought to have diverged some five million to four million years ago, a date that helps anchor the tree drawn from the new genetic data. The split between Neanderthal and human mitochondrial DNA, which marks the start of the split between the human and Neanderthal lineages, would have occurred between 690,000 and 550,000 years ago, the authors say, while the individual from whom all modern human mitochondrial DNA is descended would have lived 150,000 to 120,000 years ago.

Acknowledging the uncertainty in these dates, the authors say they show at least that Neanderthal lineage is four times as old as the human lineage, as measured by mitochondrial DNA.

The Munich team's report ranges over the three treacherous fields of paleoanthropology, ancient DNA and the genetics of human evolution. Dr. Svante Paabo has criticized many claims about ancient DNA and sought to lay out his methods with care. Still, the interpretation of the Neanderthal mitochondrial data may be open to debate on the dating.

"Deriving these dates involves making a lot of supposition about the neutrality of the mitochondrial genome and the speed of accretion of new changes," said Dr. Ian Tattersall, a paleontologist at the American Museum of Natural History in New York City.

But Dr. Ruvolo said the Munich team's methods seemed sound and its interpretation of the data was likely to be accepted. "There could be minor quibbles over the dates but the overall properties of the tree won't change," she said.

—NICHOLAS WADE, July 1997